PRAISE FOR PSYCHEDELIC JUSTICE

"Labate and Cavnar have done it again: an excellent, timely anthology that addresses crucial issues in the psychedelic community of social equity, the globalization of psychedelic substances and culture, and our shared responsibility to prevent the extinction of these plants and animals."

— Julie Holland, MD
Author of *Good Chemistry: The Science of Connection from Soul to Psychedelics* (Harper Wave, 2020)

"A powerful and thought provoking collection of essays that confront our colonial and patriarchal collective shadow. Deeply informative and challenging, the way thinking ought to be these days, as we are taking giant leaps towards psychedelic mainstreaming."

— Maria Papaspyrou
Co-editor of *Psychedelics and Psychotherapy: The Healing Potential of Expanded States and Psychedelic Mysteries of the Feminine*

"Questions about whether and how psychedelics can lead to a better world have abounded for decades in the West, but it's notoriously hard to translate profound psychedelic experiences of unity, transcendence, and love into values-driven action in our everyday consensus reality. Dr. Labate and Dr. Cavnar's new edited book *Psychedelic Justice* highlights many of the challenges we face in navigating diversity, equity, access, and ethics in this current psychedelic renaissance. These are not easy topics, but by addressing spiritual bypassing and engaging in mutually respectful dialogue, we can raise voices that too often are silenced. There's enough room for all of us to be included—in fact, it benefits everyone to ensure that's so."

— Kile Ortigo, PhD
Author of *Beyond the Narrow Life: A Guide for Psychedelic Integration and Existential Exploration*

"*Psychedelic Justice* is an inspiring and important collection of essays that ask the hard questions the psychedelic community needs to grapple with to move forward with integrity."

— Michelle Janikian
Author of *Your Psilocybin Mushroom Compani*

"This volume should be necessary reading for anyone interested in psychedelics or is in some way part of the so-called psychedelic renaissance. It brings together diverse voices that do a marvelous job highlighting the difficult conversations within the community. Read with an open mind and prepare to be humbled."

— Evgenia Fotiou, PhD., Cultural Anthropologist

"For those at the crest of the wave, the once illicit, now burgeoning emergent field of psychedelic research and treatments, offers immense and rich possibilities. Who is included, who has safe access, who has the power and privilege of participating, dispensing, and using psychedelics, are crucial issues and questions that must be brought to the fore. For People of the Global Majority, POC, BIPOC, and Queer communities, wondering where you fit in, in the field of psychedelics, and for all others who care about social justice in healing, the dynamic and diverse voices presented in the beautifully written, Chacruna Anthology, *Psychedelic Justice,* provide a vitally important, cultural and historical resource that passionately and thoughtfully explores these issues."

— Licia Sky
Co-founder and CEO of the Trauma Research Foundation

"Sharp, original, and insightful! *Psychedelic Justice* presents a series of unique and informed perspectives that are truly engaged with cultural diversity and reciprocity. Every chapter is a breath of fresh air that embraces an astonishing journey in the psychedelic landscape."

— Osiris González
Post-doctoral Researcher in Cognitive Freedom and Psychedelic Humanities

"Finally I can see myself, my ancestors, my children reflected in a text about psychedelics! This is a necessary book for anyone in the field to add to their scholarly collection, particularly Black and Brown folx in the psychedelic space who find themselves often missing from the pages of popular publications about the 'psychedelic renaissance.' Thanks to Chacruna for this offering, a full picture of what these times and these medicines mean for all of us and so beautifully capturing these 'missing voices' by giving them a platform to speak with this collection."

— Courtney Watson, LMFT
Owner of Doorway Therapuetic Services

PSYCHEDELIC JUSTICE

Toward a Diverse and Equitable Psychedelic Culture

EDITED BY

Beatriz Caiuby Labate
and Clancy Cavnar

SYNERGETICPRESS

Synergetic Press 1 Bluebird Court, Santa Fe, NM 87508 |
24 Old Gloucester St. London, WC1N 3AL England

Library of Congress Cataloging-in-Publication Data is available.

ISBN 9780907791850 (paperback)
ISBN 9780907791867 (ebook)

Cover design by Brad Greene
Book design by Howie Severson
Managing Editor: Amanda Müller
Printed in Canada by Marquis

Table of Contents

SECTION 2: PERSPECTIVES ON CULTURAL APPROPRIATION, COLONIALISM, AND GLOBALIZATION OF PLANT MEDICINES

SECTION 3: PSYCHEDELICS AND WESTERN CULTURE

SECTION 4: QUEER

Note from the Editors

The Chacruna Institute for Psychedelic Plant Medicines was co-founded by Brazilian anthropologist Bia Labate, PhD, and American psychologist Clancy Cavnar, PsyD, in San Francisco in 2017. We provide public education about psychedelic plant medicines and promote a bridge between the ceremonial use of sacred plants and psychedelic science. As psychedelics go mainstream, we curate critical conversations about controversial and marginalized topics in the space. We also promote access, inclusion, and diversity by uplifting the voices of women, queer people, Indigenous peoples, people of color, and the Global South in the field of psychedelic science.

We hope you enjoy this wonderful collection of articles, which were published originally in the Chacruna Chronicles online. We also hope this book stimulates critical thinking about access, inclusion, diversity, cultural appropriation, conservation, regulation, and implications of the emergent commodification of psychedelics.

At Chacruna, we plan to continue working to help rewrite the narrative the psychedelic field tells itself. Stay tuned!

Additionally, we would like to thank our friends at Synergetic Press for their dedicated work, all the contributors to this anthology, and the remainder of the Chacruna Team.

Beatriz C. Labate & Clancy Cavnar
San Francisco, June 2021

Preface

ERIKA DYCK, PHD

The concept of social justice has been with us for over a hundred years. It initially emerged in the 19th century among disgruntled workers in Europe, frustrated that they provided labor, sweat, raw energy but did not get to share in the spoils of industrialization. They relinquished their bodies and minds to pursue a better life and future, but in return, felt the growing gulf of income inequality and the sting of discriminatory systems that restricted their access to education, health care, and legal power. Famously, the grievances experienced by an expanding working class around the globe erupted in demonstrations of collective action that generated an ideological commitment to equality.

In the 21st century, social justice has expanded beyond its working class roots and evolved to embrace a more diverse way of understanding marginalization, discrimination, and inequalities that go beyond wealth disparities alone. Today's social justice advocates recognize a more complex set of structures and systems that produce inequalities by addressing sexism, racism, ableism, capitalism, and colonialism alongside homophobia, transphobia, climate change denial, sexual abuse, and ecological destruction. Social justice advocates confront and reject a set of systems and attitudes that perpetuate inequalities.

Despite the change in tone from the historical workers' revolts, collective action remains at the forefront of justice work. But collective action is complicated when we appreciate the intersectionality of injustice in today's world. That is to say, how do we prioritize a growing list of inequities, especially when drawing focus away from one to concentrate on another can cause divisions? Are wealth inequalities more of a priority than environmental concerns, or should we first focus on systemic racism? These goals are not necessarily oppositional, but they can divert energy a very different chemical reaction, one focused on numbing the pain and trauma caused by discrimination leveled at the laboring classes. For Marx, religion created a distraction from that pain that fuelled campaigns for justice. But psychedelic drugs are not opiates for the masses—they fulfill a very different role. Rather than being a distraction, psychedelics provide a way to connect pain and healing through conscious engagement. They are a means to an end, not an end in themselves. Psychedelics alone will not vaccinate us against a pandemic of isms.

The legacy of connections, collective action, and healing from psychedelics has a history stretching far beyond the coining of the word. As contributors in this volume show us, Indigenous ceremonies with plant medicines have a long tradition of linking people with plants in the confrontation of pain.

We can experience pain on an individual level, like overcoming an unhealthy relationship with opiates. We can also experience pain collectively, such as the despair felt by many when Jair Bolsonaro was elected as President of Brazil. His policies have wreaked havoc on the Amazon rainforest and Indigenous communities, making us feel like the clock of social and environmental progress is regressing.

Sometimes that pain stems from injustice, whether that be feelings of displacement from one's home, disenfranchisement from one's community, disorder from one's own body and mind, or disunity from one's social environment. Contributions in this collection introduce us to these topics by unpacking the history of sexism, patriarchy, homophobia, and binary thinking that causes harm—issues especially pervasive in medical and legal fields. We also look at ways to rethink trauma and racism.

Psychedelics nourish these topics with intellectual energy to visit sites of trauma with open minds and hearts. These contributors give us constructive guidelines for respectfully inviting Indigenous participants into conversations without tokenism, offer cautionary advice about the risks and rewards of bringing psychedelics "out of the shadows" and into the marketplace, and teach us how to recognize sexual abuse in intimate healing settings. Taken as a whole, the authors embrace the spirit that psychedelics are indeed a project of fascination; thinking with and about psychedelics allows for greater investments in changing our circumstances—as individuals and as a collective.

This Chacruna anthology unites authors who recognize the power of psychedelics to change the way we think. It is not just that we need to update the browser on our search engine, but rather that we need to unplug the computer altogether and look instead to different sources of information, be it from the stars, plants, or people in our communities--especially those who are different from us.

Social justice advocates ask us to think critically about the structures of power in our world that perpetuate inequalities, and psychedelics offer us an intellectual passport to visit these sites of injustice, to understand the violence perpetuated by discrimination, and to generate the courage to change history. It is then up to us all to integrate the insights we gain from our experiences with psychedelics to create a more just, tolerant, and supportive world. The authors in this collection give us a lot of tools to begin the process of integration.

Why Black People Should Embrace Psychedelic Healing: Reclaiming a Cultural Birthright

MONNICA T. WILLIAMS, PHD, ABPP

Exclusion, oppression, mass incarceration, and research abuses have become barriers to the ancient wisdom of healing processes we need and deserve.

BLACK PEOPLE LIVING IN WESTERN CULTURE OFTEN FEEL THE IMPACTS OF racialization, cultural trauma, and racism. Oppression and inequality can contribute to psychic distress at individual and collective levels, and this is enhanced and compounded for people who exist at the intersection of multiple marginalized identities.[1]

Recently, there has been much discussion about the use of psychedelic medicines to address a broad array of mental health conditions, including depression, post-traumatic stress disorder, addiction, end-of-life anxiety, and other conditions.[2] In fact, several top universities have even started centers for psychedelic research—including Johns Hopkins University, Imperial College in London, and the University of Toronto—and educational programs are now being developed to train clinicians to deliver psychedelic-assisted therapies.

There seems to be a vast potential for psychedelics to help heal many types of psychological difficulties, including racial trauma—a form of post-traumatic stress disorder brought on by ongoing experiences of racism.[3] But this potential is unrealized, especially for Black people who are part of the African diaspora. As an African American clinical psychologist and researcher, it has become evident that we have not been meaningfully included as research participants or researchers in scientific studies, and our voices have not been part of the conversation as these medicines move into mainstream mental health care.[4]

In my own community, African Americans tend to be wary of psychedelics due in part to the stigma attached to illicit substances. Back in the 1980s, the Black church was motivated to find solutions to the widespread use of crack, and so it aligned itself with the Reagan-era "War on Drugs" as a potential solution. But, at every level, the criminal justice system is heavily biased against Black people compared to Whites, from racial profiling to arrests, sentencing, and onward.[5] So, the War on Drugs became an excuse for the mass incarceration of Black Americans accused of drug infractions.

In addition, there is evidence that the risks of early era psychedelic research (1950–1985) unduly rested on the backs of African Americans and other vulnerable populations. In my lab, we have been examining these early studies, comparing the treatment received by White research subjects to what was experienced by people of color. The Addiction Research Center (ARC) in Lexington, KY, run by Dr. Harris Isbell, shared the campus with the Federal Bureau of Prisons. The research subjects were inmates—a third White, a third "Negro," and a third "Mexican." Many have heard of the Tuskegee Study of Untreated Syphilis, but few know about the facility dubbed the "Narco Farm."

One study describes two groups that received LSD. The first was a group of "Negro" males convicted on drug charges who were recruited from prison and provided coercive incentives (heroin) to participate in dubious LSD experiments. The comparison group was professional White people at Cold Spring Harbor, living freely, who were not coerced but given LSD in the lead researcher's home "under social conditions designed to reduce anxiety."[6] Knowing the profound influence of (mind)set, setting, and intention, it is certain that people in these two groups had very different experiences.

There were over 500 published studies that came out of ARC from 1935–1975, testing the limits of human tolerance for psychedelics, opiates, and amphetamines on prisoners.[7] Dr. Isbell's studies included dangerously high and prolonged doses of LSD on his subjects.[8] In the 1970s, ARC moved to Baltimore and became the National Institute on Drug Abuse (NIDA) after a nationwide ban on the use of federal prisoners as research subjects. But NIDA has never renounced these studies, although they violated well-established guidelines for the ethical conduct of biomedical research (e.g., Nuremberg Code, Declaration of Helsinki, Belmont Report).

Between the barbaric research conducted on captive African Americans and mass incarceration justified by the War on Drugs, it is unsurprising that so many Black people are uninterested in psychedelic medicine today. They may not know the details of the crimes committed against us, but the cultural memory remains. I hope that somewhere in the recesses of our cultural consciousness, there still exists some memory of the valuable psychedelic traditions cultivated by our African ancestors and nature's many gifts that have been lost somewhere between North America and the Middle Passage.

The city of Oakland, just east of San Francisco, has traditionally been an African American enclave, though it is now experiencing shifting demographics and gentrification. Nicolle Greenheart is a trained facilitator for Sacred Garden Community in Oakland, where she works with diverse communities within the ceremonial entheogenic healing space. Nicolle is also a co-founder of the Decriminalize Nature movement, the non-profit organization responsible for decriminalizing psychedelic plant medicines in Oakland.

I had the pleasure of meeting Nicolle at a ceremonial retreat in the Bay Area, and I was impressed by her dedication and commitment to psychedelic healing. She also understood the reluctance of Black people to engage in the work. "I could completely relate to the belief that psychedelics were just White people hippie drugs tied to the wild and crazy 60s," she noted. "They definitely held a stigma for me. But then I did my research and started working with the medicine and discovered for myself the life-transforming and healing benefits of it. It also immediately became clear to me that my people could benefit from this kind of therapy but are being left out of the psychedelic renaissance that is unfolding. I know what it's like to be the only Black person in the room talking about the healing power of plant medicine and wondering how I can get more of us not only in the room but also truly benefiting from the medicine in a way that feels safe and honors our own ancestral roots."

There has been much written about the Indigenous use of plant medicines from Mexico and South America, but psychedelics have been used across cultures and eras. Psychedelics were used in Biblical times to anoint priests and kings,[9] and they have also been used for thousands of years in African cultures. During slavery, Yoruba women from West Africa performed healing roles using their knowledge of plant medicines derived from Africa. During current times in Ethiopia, all plants are believed to possess some degree of medicinal usefulness, and medicinal plants occupy a central place in their traditional healthcare system. Many plants are bred and conserved in sacred community gardens, and families also keep small home gardens. This includes an array of flora for medicinal purposes and important psychoactive plant medicines for psychological and even spiritual problems.[10]

There has been much interest in the West African shrub iboga, a powerful psychoactive plant medicine that is the source of ibogaine. It has been used for centuries in healing ceremonies and cultural rites by traditional communities in West Africa among members of the Bwiti religion in Gabon, Cameroon, Equatorial Guinea, and the Congo.[11] In the West, it was discovered that ibogaine can significantly reduce withdrawal symptoms from opiate dependency and eliminate cravings. This has resulted in several noteworthy clinical trials and thriving international ibogaine treatment centers for opiate addiction—one of Africa's many gifts to the world.

In southern Africa, there is widespread reliance on *ubulawu* as psychoactive spiritual medicine used by Indigenous people groups, such as the Xhosa and Zulu, to communicate with their ancestors and treat mental disturbances.[12] Ubulawu, an ancient African plant medicine, is composed of the roots of several potent plants that are ground and made into a cold water infusion, churned to produce a healing foam. Scientists in the West are now trying to ascertain how this traditional African medicine can open self-knowledge and intuitive capacity. The Bushmen of Dobe in Botswana use the hallucinogenic plant kwashi (*Pancratium trianthum*) for spiritual and healing purposes.[13] It is not known how many psychedelic plants can be found in Africa, but as of 2002, over 300 plants with psychoactive uses had been identified in South Africa alone, many with psychedelic properties.[14]

There is such a rich tradition of plant medicines in Africa that it is clear Black people have benefited from psychedelic plant medicines for a very long time. At my mental health clinic in Connecticut, we have started providing culturally-informed psychedelic-assisted psychotherapy as a means of treating racial trauma.[15] We don't have access to any of our ancestral plant medicines, and most psychedelic chemicals are not yet available outside of clinical trials. But we do have ketamine, which has been shown effective for certain mental health problems, and it can be used effectively for psychedelic psychotherapy.[16] I have spoken to a multitude of Black people who have been wounded by large and small blows from discrimination accumulated over a lifetime. Many are fearful of psychedelic medicines and the vulnerability that comes with being in an altered state. We know that, certainly, these treatments can be unsafe without skilled providers or caring therapists to guide them on the journey.[17] But these medicines are part of our cultural birthright, and I believe we lose more when we step back and choose not to engage. It is true that it has not always been safe for us, but I hope we can come together as a people, create our own safe spaces, and become empowered to reclaim psychedelic healing for ourselves, our loved ones, and our communities.

To this point, Nicolle Greenheart says, "Now it's part of my life's work to do what I can to help educate my people and create safe healing spaces for us, by us. Given the cultural climate we're living in these days and the historical trauma we've endured, we truly need safe spaces to heal together, in community."

Mestre Irineu: A Black Man Who Changed the History of Ayahuasca

GLAUBER LOURES DE ASSIS, PHD

YOU PROBABLY KNOW ABOUT AYAHUASCA. YOU PROBABLY LOVE THE Amazon. You probably support Black Lives Matter (at least, I hope so!). What you probably don't know is that a Black man who lived in the Amazon played a major role in the history of ayahuasca. In this short essay, I will introduce you to Raimundo Irineu Serra, who founded Santo Daime, the oldest Ayahuasca religion in Brazil.

A SLAVE'S GRANDSON GOES TO THE AMAZON

Raimundo Irineu Serra was born in 1890 in a small town in Maranhão, Brazil's poorest state, just two years after the country abolished slavery.[1] The grandson of enslaved Black people, he grew up in poverty and never had the opportunity to attend school. Coming from a background filled with adversity and inequality, none of his childhood friends could have imagined that Irineu Serra would overcome such difficulties and structural racism, but he did. Nonetheless, he suffered many challenges and struggles along the way, struggles that are connected to the ones Black people in America and worldwide endure today.

Seeking a better life, Irineu left Maranhão in Northeast Brazil and made the long journey to the Amazon, to the territory now known as Acre. He arrived there in 1912, motivated by the government's promises that rubber tappers could earn a prosperous living in the forest. That promise was a lie, of course. The rubber economy was built upon the merciless exploitation of workers who usually lived in extremely poor conditions. To make matters worse, due to the biopiracy of rubber seedlings, Brazil lost its monopoly on rubber production to the plantation system of British colonies in Malaysia, which caused a socioeconomic collapse in the region.[2]

Thus, thousands of immigrants who, like Irineu Serra, dreamed of having a better life in the Amazon were utterly abandoned. It was in this context that Irineu, almost without hope, sought ayahuasca.

FROM MOON WOMAN TO BLACK MAN: INITIATION WITH AYAHUASCA

According to reports by his followers, Mestre Irineu was probably initiated into the mysteries of ayahuasca in Peru by Indigenous shamans.[3] When he took the enigmatic potion, Irineu began to have spiritual revelations that transformed his life and his understanding of the brew. According to Daimista mythology, on a clear and beautiful night, Irineu took ayahuasca and, as he looked up at the moon, he saw a beautiful and wondrous lady. She asked him, "Who do you think I am?" Amazed, Irineu looked at her and replied: "My lady, you must be a Universal Goddess!"[4] This female entity was later identified as the "Queen of the Forest," and is also understood to be a manifestation of the Virgin of the Immaculate Conception.

She then directed her disciple to go through a series of fasts and trials, after which she granted him the right to make a wish. Young Raimundo asked to become a great healer to help people and requested that she put all her healing powers into that brew. She granted him this wish, and thus Irineu Serra became Mestre Irineu.

SANTO DAIME: A CULTURAL MOSAIC

Through this encounter between a female deity, a Black man, and an ancestral magical beverage, ayahuasca was renamed "Daime." Daimistas say the word comes from a formal and somewhat archaic imperative form of the verb *dar*, "to give": *Dai-me força, dai-me amor*, "Give me strength, give me love," they chant in their ceremonies. In order to navigate the pathways of the mysterious dimensions of ayahuasca, one must adopt a humble position before its power and ask: "Give me!"

Mestre Irineu began to blend the ritual consumption of ayahuasca with elements from other spiritual traditions, including folk Catholicism, Indigenous spirituality, European esoterism, and Afro-Brazilian religions, giving rise in the 1930s to an original philosophy known today as Santo Daime. Thus, since its inception, Santo Daime has been characterized by syncretism and exchange with other spiritual practices, constituting itself as a cultural mosaic. This allowed the group to adapt to the complex and discordant context of the time and be molded by different conceptions of the world.

This account, which reveals Mestre Irineu as a shaman who built bridges between different worlds, gives us clues for understanding the presence of Christian elements in Santo Daime. However, some people, especially middle-class White populations of the Northern Hemisphere, are resistant to these Christian associations. It is commonly said that Mestre Irineu "Christianized" ayahuasca. What is left unsaid is that he also "Africanized" Christianity, giving it new meanings for poor people of the rubber plantation: a Christianity that rescued the self-esteem of racially mixed rubber tappers—a decolonial Christianity. We need only to look at Santo Daime's origin story: the Catholic Virgin of the Immaculate Conception reveals her spiritual mission to Irineu Serra, but she is also simultaneously the Queen of the Forest and the "Universal Goddess." Just as ayahuasca is a symbiotic mixture of two plants that produce a unique brew, Mestre Irineu's philosophy is an anthropophagic mixture of different cultures that produces an authentic identity.

A MUSICAL EXPERIENCE

Like many other Daimistas, I like to call Santo Daime a "musical doctrine." After he established this initial contact with the Queen of the Forest, she told Irineu that she wanted to be praised through cheerful songs. Irineu replied that he did not know how to sing, at which she ordered him to open his mouth so that she could teach him. At that moment, Irineu was inspired to sing "Lua Branca," the first hymn of Santo Daime. Since that time, the sacred texts of the religion are the musical hymns "received" by Daimistas from all over the world. Mestre Irineu's hymnbook, known as *O Cruzeiro* ("The Holy Cross"), includes 132 hymns that form the basis of the group's religious practice. Organized chronologically, these songs are considered by Daimistas as divine, spiritually inspired messages.

It is virtually impossible to understand Santo Daime without considering its musical elements. Music is not merely an aspect of Santo Daime practice, rather, it creates the enchanted universe of "The Doctrine." The soundscapes of the ceremonies are capable of carrying away and bringing people back from their visionary journeys and are a potent illustration of the symbiosis between psychedelics, culture, spirituality, and the body.[5] In Santo Daime ceremonies, music is so important that people not only listen to hymns but also sing and dance to them for as long as 12 hours, depending on the ceremony. The Daimista is a dancer by nature! Santo Daime ceremonies, referred to as *trabalhos* ("works"), are, most essentially, musical experiences.

FROM JAIL TO JERUSALEM

As a Black man who dared to cultivate an anti-hegemonic spirituality, Mestre Irineu suffered persecution and was strongly stigmatized for a long time. In his early days in the Amazon, he was the target of police actions against "witchcraft." He was even shot by the police. For these reasons, he moved from the Amazon interior to the state capital, Rio Branco. But that didn't make things better. Irineu Serra, who also caught everyone's attention due to his height—he was about 7 feet tall—was accused of charlatanism and quackery in Rio Branco and even went to prison in 1942 after a siege of his home by 40 police officers.[6] This incident motivated the Master to move once again, this time for good. And yet, his persecution led Irineu to develop diplomatic skills, allowing him to move between different social circles. He established alliances with important politicians while rescuing people who lived in virtual slavery at the rubber camps. Mestre Irineu's cosmopolitical diplomacy also included plants, animals, and nature. He was able to consolidate his religious group, which grew to a few hundred people, making Irineu a regionally respected citizen.

This consolidation allowed Santo Daime, decades later, to catapult into a national and international diaspora, undertaken especially by two of his followers, Padrinho Sebastião Mota de Melo and his son, Padrinho Alfredo Gregório de Melo. Today, the movement founded by Irineu Serra is present in at least 43 countries on all inhabited continents. From Tokyo to Berlin, from New York to Jerusalem, Santo Daime is one of the most widely spread Brazilian religious movements. It has also played an important role in the regulation and legalization of ayahuasca worldwide.[7]

This is not to say that things have become easy for Daimistas. On the contrary, persecution continues globally now. In the past three decades, Daimistas have been arrested, and dozens of liters of the Daime brew have been seized worldwide. But Irineu Serra's followers are resilient—they get it from him!

THE ANTHROPOMORPHIZATION OF AYAHUASCA: IRINEU BECOMES JURAMIDAM

Mestre Raimundo Irineu Serra died on July 6, 1971. His followers believe that his death symbolizes the fulfillment of his mission on Earth; by then, his doctrine was ready. He then came to be recognized by Daimistas as "Juramidam," his spiritual name: "Imperial Master Juramidam." Ayahuasca, understood by Daimistas as a "divine being, transformed into liquid"[8] has become synonymous with Mestre Irineu himself. The brew has transformed into a person, and Irineu Serra has become a plant: Daime and Irineu merge into one. As he once said, "I am Daime and Daime is me."

Today, the tomb of Mestre Irineu is a place of international pilgrimage and an important tourist destination in the city of Rio Branco. The ayahuasca formula developed by Irineu was consecrated not only in Santo Daime but in several other religious groups as a safe method of preparation and a very special ayahuasca brew. The original community continues to function in the same place, led by Irineu Serra's widow, Madrinha Peregrina Gomes Serra, who is now 83 years old and continues to participate actively in religious life.

ROOTS

Today, ayahuasca is becoming mainstream. And yet, alongside this valorization of traditional knowledge, this movement has also commercialized the ayahuasca brew and, in many cases, disenchanted its practices. Ayahuasca practices appear to be going through a "Whitening" process, and its founding figures are being erased from history. Revitalizing the memory of historical figures such as Mestre Irineu from the Global South, the cradle of ayahuasca, is an important way to counteract this process.

Mestre Irineu is a fascinating personage in the history of ayahuasca. His life teaches us that discussions around this "drink that has incredible power"[9] are not only about psychedelics but also about diversity, environment, culture, and overcoming prejudice along with social stigma. The trajectory of Raimundo Irineu Serra reminds us, once again, to acknowledge the presence and contributions of Black people everywhere, including wherever ayahuasca is found and consumed. May this simple testimony contribute to keeping the memory of this great Black Brazilian man alive. As he tells us:

> "Here I come to my end
> I leave my narration
> To always remember
> The old Juramidam."[10]

How Women Have Been Excluded From the Field of Psychedelic Science

ERIKA DYCK, PHD AND CHACRUNA INSTITUTE

CHACRUNA INTERVIEWS HISTORIAN ERIKA DYCK, WHO WAS INVITED TO join us on the Women and Psychedelics Forum,[1] but whose schedule did not allow her to attend. The interview addresses some of the historical attitudes about women in psychedelics based on her research. She has studied an earlier generation of psychedelic science from the 1950s and 1960s. Dyck is the author of *Psychedelic Psychiatry: LSD from Clinic to Campus*, and editor of *A Culture's Catalyst: Historical Encounters with Peyote and the Native American Church in Canada*, and the forthcoming book, *Psychedelic Prophets: The Letters of Aldous Huxley and Humphry Osmond.*

1. What are women's historical contributions to the field of psychedelic science?

In my research into 1950s experiments, women were central but often invisible contributors. A large number, maybe a majority, of early experimenters had their first experiences with their wives. Some husbands confessed to me later that they were scared and wanted to experience something with the person they trusted most. Wives helped them write up their experiential reports and often sat with other volunteers in subsequent sessions as experienced sitters or guides, even before that terminology became more commonplace. In my research, however, none of these wives ever appeared on a published paper or report, and some were not even identified by name. Mrs. Al Hubbard, for example, seemed to have sat in on at least as many sessions (maybe more), but her first name is not even recorded, yet Captain Al Hubbard became infamous for his role in this history. Maria Huxley, Aldous Huxley's first wife, was a major influence on him and pushed Aldous to more deeply explore parapsychology and Indigenous rituals, largely through her own network of women.

2. Can you say more about those "wives" who were so critical to that early research?

It is probably difficult today to fully appreciate what it might have been like to live as the wife of one of the early psychedelic enthusiasts. The 1950s are not held up as a moment of progressiveness for women in the workplace or for finding (or being discovered for) their public voices. It seemed that the 1960s generation overshadowed any such strides that 1950s women might have presented. But, imagine a time when a single-family income was more the norm and where the family income was indeed that. There seemed to be a sense of partnership and shared responsibility for the husband's success, in this case, his research. I don't want to romanticize or overshadow some of the obvious pitfalls with this logic, but I think it is important for considering how men like Aldous Huxley, Al Hubbard, Humphry Osmond, and others relied on their wives for providing absolutely critical feedback on their work, almost as an expectation.

This is not to cast these men as monsters, but it is an attempt to contextualize the husband and wife partnership in the 1950s, and especially in a moment where these men felt quite vulnerable. It also, I think, helps us to recognize the vital contributions that women have been making to this field for a long time.

Each of these men had their first "psychedelic" experience with their wife. Each later confessed to being nervous and apprehensive but calmed by the presence of their loving partners. Some of these women later recorded their own feelings too. Rose Hoffer (Abram Hoffer's wife) remembered mostly feeling sick and worried that this would ruin the experiment when it turned out that several other people had also felt nauseous but were uncomfortable admitting it. Ellen (Matthew Huxley's wife, son of Aldous Huxley) coaxed her husband through a difficult first experience and later took a real interest in recording their subsequent reflections about that moment as a way to catalog the insights they gained. Indeed, it was she who kept up the correspondence. Mary Agnew, the wife of psychologist Neil Agnew, remembers agreeing to try LSD with her husband to ease him into the experience and how he invited her to help guide subsequent students and volunteers in his lab after that day. He told me, in his 90's, that she was the best guide there was.

I am only speculating, but it strikes me that these women may have been instrumental in helping their husbands feel more confident about their work. Some clearly helped to articulate the experience, lending credibility to the vocabulary of the description and helping to create new frameworks for understanding as they talked with each other and formed their own networks of support for one another, as well as for the men. Today, we might call this emotional labor, yet we still struggle with how to evaluate it.

3. How would the field of psychedelic science be impacted if there were more women?

First, everything would be better if there was more diversity of experience and understanding. Like other historical examples of medical experimentation, psychedelics were and still are largely dominated by men. I think this also lends itself to a culture of bravado, machismo, and a style of confidence in assertions that we don't necessarily see from the women who are routinely involved in the experiments or counseling sessions, though rarely as PIs. My historical research suggests that women were almost always involved in the counseling sessions, recruitment, etc., but are very rarely identified in the published work. The legacy of that history continues to distort our understanding of who does the work and what kind of work is valued.

4. What is the structure of the majority of psychedelic research institutions and non-profits? Are they composed of men or women? What about their boards?

In Canada, our Minister of Science, Kristy Duncan, recently scolded post-secondary institutions for the lack of diversity in research chairs. She insisted that universities make up at least 30% of major-funded research chairs with either women, people with disabilities, or visible minorities. My major university had 17% diversity, according to these criteria. We all need to do better. I am not convinced that being a White male makes one more capable of original and innovative research. I am convinced that being White and male has had historical privileges associated with it that are now assumed rather than assessed.

5. How does women's history of oppression inform their approach to psychedelic science?

Historically, women were first legally prohibited and then culturally excluded from public drinking establishments. This led to a serious distortion in the number of cases of women diagnosed with alcoholism who entered the 1950s generation of psychedelic trials for alcoholism. It was not that women weren't there, but they received different kinds of therapy and were more often treated for depression or anxiety than for alcoholism. At that time, some of the conventional wisdom suggested that psychedelics offered an appropriate male therapy because it was potentially scary or even destabilizing. Practitioners even suggested that this was better suited for men who were attracted to bars where they could fight each other and seduce women. It took a lot of courage to go into an LSD trial, especially in an era where most psychiatry and psychotherapy was done without any psychoactive substances at all.

Despite the outcomes and the acknowledgment that this perspective was overblown, much of the early research concentrated on an ethos of bravery or a daring spirit that was explicitly designed to target men.

6. Your new book talks about letters between Huxley and Osmond, two men. How do women figure there?

Yes, *Psychedelic Prophets* is a critical edition of the complete letters of correspondence between writer Aldous Huxley and psychiatrist Humphry Osmond: the two men who coined the word "psychedelic." It is true that this book falls into the trap of celebrating the male pioneers of psychedelics, but I am proud to say that the editorial team was also quite conscientious about capturing the additional letters, where possible, or notes, in some cases, written by the women in their lives. Their wives, Jane Osmond, Maria Huxley, and later Laura Huxley, Osmond's young daughters Euphemia and Helen, and Huxley's daughter-in-law Ellen, not only appear in references but are woven into the lives of these men as they write about their research and of their families. Indeed, I hope that these letters might give us a glimpse into that historical moment, a mixture of curiosity and trepidation about what psychedelics might do for humanity. Some of that trepidation comes through in how they talk to the wise women in their lives, wives, as well as others, like parapsychologist Eileen Garret, who features prominently in their letters. Despite the published names, they demonstrated a kind of deference and respect to their wives as confidantes and wise women who played an instrumental role in their work. I hope that will come across to readers too.

7. Where do you see the future of psychedelic science headed? Or where would you like to see it headed towards?

As a historian, I suppose it is cliché to say that I hope we learn from our past, but in this case, I think some of the lessons are clear. The lack of diversity in science is not just a problem for those of us interested in psychedelics, but in some ways, the stakes might be even higher. Many of the guiding principles behind psychedelic therapy embrace subjectivity and encourage us to tolerate diversity of expression and experience. We need to then lead by example as a community of researchers who can bring this philosophy into practice as we train, research, and disseminate our findings with diversity in mind. We know that things like emotional labor are a critical part of this enterprise, that binaries are indeed more fluid, and the way we do or don't feel secure in a space has a significant impact on how we experience mental health. I guess that means I would like to see members of the psychedelic science community champion these values and help set an example of how mental health care can be done.

8. Anything else you wish to say?

I wish I could go to the Women and Psychedelic Forum! I hope it is a wonderful meeting. Keep up the good work!

When Feminism Functions as White Supremacy: How White Feminists Oppress Black Women

MONNICA T. WILLIAMS, PHD, ABPP

RECENTLY, I PARTICIPATED IN A PANEL DISCUSSION AT A PSYCHOLOGICAL conference in Washington DC. The name of the panel was "Being a Mentor in the Era of #MeToo," which seemed extremely timely given the spate of powerful people recently felled by revelations of sexual wrongdoing. As the only person of color on the panel, I wanted to make sure to mention some of the issues around this movement that singularly impacted people of color. The #MeToo movement included responsibility for privileged women to stop victimizing men of color, many of whom have been lynched, beaten, incarcerated, and electrocuted on the false whisper of a White woman. I had recently supported a member of our research team who ended up in such a predicament, and so this point felt particularly real for me at the moment.

As I shared my observations and convictions, I was abruptly cut off by the moderator, a senior academic White feminist. The mic was taken from me, a woman of color, and handed to a White woman who had her own personal story of oppression to share. I was silenced, my uncomfortable words forgotten, and I became irrelevant. At the podium, the moderator affirmed her lifelong commitment to feminist ideals while I was shamed into submission. This is White feminism.

UNDERSTANDING WHITE FEMINISM

True feminism has the power to transform society, but too often, what is advanced as feminism is actually White supremacy in disguise—a counterfeit we sometimes call White feminism. White feminism exists to promote the comfort and safety of middle-class and affluent White women.

At its core, it is a racist ideology that claims to speak for all women while ignoring the needs of women of color and suppressing our voices when our agendas and priorities don't align. It recognizes the voice of women of color only to further its own aims and appear inclusive.[1] Its organizational representations fail to properly address racial and economic intersectionality in experiences of sexism. It rejects the idea that women can oppress others who are disempowered and, in doing so, replicates the harmful unacknowledged social dynamic of the primacy of well-educated White voices.

As a researcher and educator, and I have seen White Feminism at work in nearly every facet of my career. My current research areas include mental health disparities, racism and discrimination, trauma, obsessive-compulsive disorder, and the emerging field of psychedelic medicine. In the psychedelic world, I am repeatedly expected to give talks nationwide at my own expense. And, even though I have conducted multiple research studies, written books, and published dozens of articles in top journals, I find it interesting that I have had virtually no White American women faculty collaborators. I have a few White women friends, many supporters, several junior colleagues, but collaborators? None. Those same women who applaud the advances of women in academia will not collaborate, share mutual resources, or celebrate the achievements of a Black woman. Although I recently followed the drinking gourd North to a Canadian university, during my tenure at American schools, I found that White female graduate students and trainees chafed against having to answer to a Black woman, and female administrators frequently ignored or undermined my needs.

When I was invited by White female colleagues to a Women's March, I made an excuse not to go. I didn't want to be a token for a group that doesn't seem invested in women like me. Since that time, I've spoken to many other women of color who felt the same way but didn't know how to put words to their feelings. So, to that end, I am putting words to it now—and facts and figures too. I have compiled some examples of White Feminism and how it misses the needs of Black women and other disempowered women.

Black women are dying in childbirth. Maternal mortality for Black women is four times the rate of White women, and these rates remain high even for middle- and upper-class Black Women.

Even though the US spends more on childbirth than anywhere else, it's safer for a Black woman to have a baby in sub-Saharan Africa than in a modern hospital in Arkansas. Rates have been climbing for all women since 2000, and, bizarrely, no one seems to care.

It fails to fight for paid maternity leave since perhaps White women can afford to stay at home with their babies without a paycheck. We lag far behind most developed nations when it comes to maternity leave. The Trump administration has been interested in making modest gains in this area (six weeks guaranteed paid leave), with nary a peep of support from anyone on the right or left. Those on the right are concerned about the cost, while those on the left criticize the plan's shortcomings. It's a step in the right direction, and it is shameful and outrageous that feminists are not clamoring for this low-hanging fruit.

It appropriated #MeToo while failing to acknowledge how White women have historically and currently used their sexuality to oppress men of color. Many Black men sit in prison on false rape charges or face great social loss due to such accusations. When White women cry rape, our society mobilizes to punish the targets, guilty or not, to protect White women's virtue. The tragic case of Emmett Till and the Tulsa Oklahoma Race Massacre are catastrophic examples of this.

Emmett Till, a 14-year-old African American boy, was lynched for purportedly propositioning a White woman in a store. His accuser, Carolyn Bryant, recanted decades later, but it was too late for Emmett. And, in one of the largest acts of domestic terrorism ever, the prosperous Black community of Greenwood, Oklahoma, was burned down and firebombed after a White woman claimed a Black man frightened her on an elevator.

It thinks it's fine to take paternalistic control over Black women and girls' sexual choices. Pregnancy and sexuality are socially constructed as problematic and managed, controlled, and regulated depending on social status.[2] Black women and girls are disproportionately given birth control injections, implants, and the intrauterine device (IUDs) over options they can control themselves.[3] I remember very clearly after the birth of my oldest daughter, a nurse barging into my hospital room while I was half-asleep recovering from labor. She insisted I start taking progestin contraceptives and then stormed out when I didn't give in. Not only did she ignore my concerns, but this conversation should have happened while I was awake and in a place where we could have a careful discussion about my needs and choices.

It regards Black women's children as disposable. Black women's babies die at twice the rate of White women's,[4] and these rates remain the same when controlling for income and education.

Reviewing infant mortality globally, babies born in countries like Panama fare better than Black babies born in the US. Additionally, Black and Indigenous children are taken from their mothers at higher rates than from White mothers. Black children continue to be targeted throughout their lives until they are old enough to fill prisons, with Yale researchers documenting teachers racially profiling Black children starting in preschool.[5]

It has no clue how punishing and patronizing health care can be for women of color. Research shows that doctors have biases against Black patients and are less likely to engage in cooperative, patient-centered care. Doctors speak faster, dominate conversations, end visits sooner, and display fewer positive cues and less warmth, translating to lower quality health care for people of color.[6] There is almost no quantitative research on how this is uniquely experienced by women of color, perhaps underscoring what a low priority the quality of our medical interactions is to researchers and funders. But we do know that breast cancer and cervical cancer mortality is 40% percent higher in Black women than in White women. And Black people are routinely under-treated for pain, under-treated for anxiety and depression, and over-diagnosed with psychosis.

It opposes regulation to make abortion safer since they'll never need to use a low-income clinic. This did not work out so well for women of color in Philadelphia. When I was a junior faculty at the University of Pennsylvania, I walked past Dr. Kermit Gosnell's little shop of horrors daily, with no idea what was occurring inside. He performed from four to five illegal late-term abortions per week, while unqualified (even teenage) staff provided anesthesia in filthy conditions. Despite dozens of complaints, injuries, and life-threatening emergencies, no hospitals made the required reports to state health agencies, and state health agencies refused to investigate the reports that they did receive until after two women died. How did this happen? In 1993, the Pennsylvania Department of Health decided to stop inspecting abortion clinics at all. Where is all the feminist outrage over such deplorable oversight and those women's tragic and preventable deaths? A grand jury concluded that no one acted sooner because "the women in question were poor and of color, because the victims were infants without identities, and because the subject was the political football of abortion." This is a particularly troubling example of how White feminism builds its agenda atop our injured and dead bodies.

It ignores the problem of forced and coerced sterilization because disempowered Black, Brown, Indigenous women, and inmates are targeted in this eugenicist manner, but not "good White women."[7] I have experienced a small piece of this coercion myself. While in the process of giving birth, I was offered sterilization by the obstetrician, and not once, but three different times during three different deliveries. My last delivery was long and difficult because my water broke five and a half weeks early, and I endured prolonged labor as the baby was not ready to emerge. I was vomiting and having semi-conscious nightmares from the anesthetic while worried about my baby, and I don't even remember if I agreed to the sterilization or not. This did not happen in a prison, and I was not a low-income single mother. I was a faculty member at the University of Pennsylvania, giving birth at Bryn Mawr Hospital, with my husband beside me. Imagine how much harder this would be for a woman of color with far fewer resources.

It ignores oppression inflicted by the criminal justice system. Consider the shocking facts that 60% of women in jail have not been convicted of a crime and are awaiting trial, 80% of women in jail are single mothers, and two-thirds are women of color. One-third are suffering from major mental illness. Incarceration rates are increasing; Hispanic women are incarcerated at nearly twice the rate of White women, and Black women are incarcerated at four times the rate of White women. Health care is inadequate for incarcerated women, and, as noted, their reproductive rights are routinely violated. I can only assume this problem is not prioritized because White feminists don't worry about going to jail, as they can afford good legal representation and bail and are not targeted by law enforcement. Further, when

White women are assaulted and call the police, the police will almost always help them. For Black women, this is not always true (and has not been true for me).

It doesn't care that Indigenous women are disappearing and no one is looking for them. There is a really disturbing pattern of missing and murdered Native American women and girls throughout North America, and families can't get authorities to help find them. Apparently, these especially vulnerable women can be abused and murdered with no consequences at all.

It is extremely fragile around issues of race. White feminism weaponizes White women's tears in order to silence observations of racism.[8] White women merely need to cry when called out on racism and hypocrisy to villainize people of color, thereby centering the conversation on the emotional well-being of White women rather than harms done.[9] Further, it hides its White privilege behind female or LGBTQ+ oppression, as experiences of sexism and homophobia become an excuse to avoid working on their personal racial biases.

SUMMARY

It's hard to muster up the energy to fight issues like the infamous wage gap when so many of my amazing sisters of color can't get decent medical care, our babies are dying at rates typical of developing countries, our partners are sitting in jail for no good reason at all, and we are all traumatized from living in a racist society. These are my priorities, and if you care about all women, these should be your priorities too.

White feminism is so distasteful that many women of color want nothing to do with the feminist label, period. Alice Walker devised the term womanism to define her love of Black womanhood and a commitment to improved lives for all people oppressed due to race or class. White feminism is part of our oppression. When it comes to feminism in America, women of color simply don't have a seat at the table, and when we do, it's for the illusion of inclusivity and not because our differing perspectives have value. This is called tokenism: including someone from our demographic to keep up appearances when, in fact, we have no say or power.

And when we are "heard," our voices and stories are (mis)used to further White feminist goals.

A few years ago, I declined an invitation to give a keynote address at a major national feminist conference. It is tempting to think I could have used this platform to enlighten my White feminist sisters, but it would be a mistake to assume they are unaware of these issues.[10] When I recently posed these problems to a White Feminist philosophy professor, she was well aware of it, and said they were frequent topics of discussion in her classes. Therefore, I can only conclude that feminist groups don't care to do anything differently, making it difficult for me to believe that the women's movement has any good intentions for Black people whatsoever. Based on the agenda of national women's organizations, it appears that White feminist leaders have sold out to the priorities of the medical industry (run by White male doctors), big pharma, and other corporate interests (more wealthy White men). White feminist leaders are, in fact, engaging in silent collusion with our oppressors. Figuratively and literally, they are bedfellows, while women of color get the shaft.

We cannot claim to be immune from racism simply because we are "good progressive liberals." Racism is embedded in both liberal and conservative procedures and ideologies, and is even at the core of our vaunted women's movement. This brand of feminism does not speak for me.

Let's call out racism wherever it lies.

Hate & Social Media in Psychedelic Spaces

BEATRIZ C. LABATE, PHD AND NICOLE T. BUCHANAN, PHD

WHAT #BLACKLIVESMATTER SOLIDARITY STATEMENTS REVEALED ABOUT THE PSYCHEDELIC COMMUNITY

Psychedelics are often celebrated for their ability to open one's heart, create a sense of oneness with nature, [1] shift political values, [2] and facilitate empathy and love for others. [3] Psychedelics show promise for healing racism-related trauma and facilitating restorative justice across conflict groups.[4] Recently, the second author urged psychedelic researchers to reimagine their work and include healing the racism that lives within perpetrators to reduce racism perpetrated against people of color.

Although there have been some efforts to create discussion around equity and access in the psychedelic community, there is also a widespread belief that the psychedelic community is welcoming, inclusive, and devoid of racism—that, once ingested, psychedelic substances make it impossible to be racist. This is a myth that many in the community tell to avoid the hard truth that the psychedelic community has the same biased beliefs that are found throughout society, including racist beliefs. Refusal to acknowledge racism within the psychedelic community encourages its spread, but an honest look at this reality can be the first step in becoming the non-racist community we claim to already be.

This piece exposes racism in the psychedelic community as revealed in social media. Several US psychedelic organizations publicly expressed support for #BlackLivesMatter (BLM) and anti-racism efforts in the wake of nationwide protests in response to the police killings of George Floyd, Breonna Taylor, Elijah McClain, and countless other Black children, women, and men. Although there were positive responses to these solidarity statements, the many hostile responses revealed an ugly truth about the racism that persists in our psychedelic community.

We reviewed Facebook, Instagram, and Twitter social media posts responding to BLM solidarity statements posted by MAPS, Chacruna, Psychedelics Today, San Francisco Psychedelic Society, Psychedelic Support, Lucid News, DoubleBlind, and Psychedelic Seminars. We coded (see summary below) social media comments that included hostile or highly critical sentiments to either the BLM movement/anti-racism initiatives themselves or the psychedelic organization's support for these initiatives.

First, it is important to note that the majority of comments applauded the public declaration of solidarity, and hostile comments to solidarity statements were much less frequent. When they occurred, others actively denounced the hostile posts and sometimes even engaged in active debate with the person who condemned the solidarity post. Although not a conclusive finding, when we reviewed the profiles of users posting hostile, negative comments, the comments appeared to be authored almost exclusively by White men—while their opposition came from men and women from a variety of racial and cultural backgrounds.

There were interesting trends apparent when comparing MAPS versus smaller organizations. Smaller, often more alternative psychedelic organizations had fewer comments and reactions to their BLM solidarity statements and those who did respond were almost always supportive. In contrast, nearly all of the hostile, negative comments were in response to the solidarity statement posted by MAPS. This is partially due to MAPS having far more followers than the smaller organizations. MAPS also appeals to a much broader audience who hold a range of alternative and mainstream beliefs. The more mainstream audience is where racist views might be more prevalent within our psychedelic community.

This table summarizes the major categories we coded in negative, hostile comments to psychedelic organizations' solidarity statements. Although these classifications have some overlap, we believe they are useful for thinking about the beliefs and motivations that fuel anti-racism/BLM opposition.

Category	Description
Anti-BLM	-specifically opposed BLM as an organization/ movement.
Divisive	-warned psychedelic organizations that they should avoid engaging in a politicized debate because doing so will divide support for these organizations and their efforts to mainstream/ legalize psychedelics.
Ideological infiltration	-opposed BLM/anti-racism efforts and deemed them unwelcome political ideologies that are infiltrating the psychedelic movement. This was one of the most common types of comments.
Spiritual bypassing	-used spiritual, new-age language to imply that anti-racist struggles are a misuse of psychedelics, that intermingling anti-racism efforts and psychedelic work denigrates the value of psychedelics.
"Race does not exist"	-argued that race is not a legitimate category for people to identify with and blamed racism on the fact that Black people identify themselves with the illegitimate category of race.
Systemic racism denial	-denied that structural racism exists in the US and attributed racist incidents, specifically the high incidence of police killings of Black people, to either rogue cops (the "a few bad apples" analogy) or dysfunctional Black communities and Black culture.
Virtue signaling	-accuse organizations of posting BLM/anti-racism solidarity statements to appear as if they care because it is a popular position (virtue signaling). These statements also made it clear that they believed BLM/anti-racism concerns were not important or worthy of attention.

Many of these categories are found throughout the wider discourse and do not directly refer to psychedelics or the psychedelic community, such as:

1. Denial that systemic racism exists, and therefore see the BLM movement as misinformed, hysterical, opportunistic, and/or masking ulterior, sinister political motives.
2. Assertions that police violence is about Black pathology or individual bad officers, not systemic, institutionalized racism, or victim-blaming that Black people/communities provoked the violence they experienced.

Because the psychedelic community often asserts that it is free from racism, here we focus on those points that specifically linked their derision of BLM/anti-racism efforts to psychedelics and psychedelic culture.

Most users posting negative, hostile comments to the BLM solidarity statements framed such statements as reflections of a creeping ideological indoctrination that is destroying the psychedelic movement and society at large. BLM and anti-racism commitments are seen as indicators of a polluting ideology that conflates identity politics, US political party politics (Democrat/Republican), Marxism, and radical leftism. This ideology is labeled as dogmatic and crippling to the free thought and libertarianism that they believe are essential components of psychedelics. Many pride themselves on resisting this hegemonic ideology, thinking critically and rationally in opposition to what they believe is the overly emotive, victimized/victimizing mentality of the closed-minded and brainwashed (e.g., those who agree with anti-racism/BLM). Many use their ostensible clarity and neutrality to argue that anti-racist rhetoric is alarmist and deceptive, or that BLM is a nefarious organization.

These beliefs are paired with accusations that psychedelic organizations are merely virtue-signaling by posting BLM solidarity statements. While making it clear that they believed anti-racism commitments were not important, they accused organizations of expressing support to appease the public (and burnish their public image) and to capitalize on their cultural capital.

Other comments warn against this perceived ideological infiltration because they believe anti-racism and BLM ideals are antithetical to the core values inherent to psychedelics, psychedelic organizations, and the psychedelic community. These beliefs promoted psychedelics as being apolitical, even elevated above politics. In reality, these assertions are deeply political in their lack of concern about racism and simply reassert a political position that they believe is universal and therefore neutral.

WE'RE ALL ONE

The BLM/anti-racist initiatives are routinely denounced as identity politics—the tendency for people to focus on political initiatives specific to their group (e.g., race, religion, gender, sexuality) rather than traditional political goals with broad appeal—and then dismissed as divisive and a barrier to real solidarity. A common argument is that race does not exist and identifying with a race or fighting racism perpetuates the illusion of race and fuels racial antagonism between individuals. This framework negates claims of racism by people of color and simultaneously blames them for racism if it does exist.

Psychedelics are positioned as being unique tools in facilitating a post-racial transcendence, implying the effects of their use are incompatible with holding racist beliefs. With this lens, anti-racism efforts are unnecessary and, therefore, associating psychedelics with Black liberation is met with suspicion. The fact that someone could use psychedelics and still believe that racism exists, particularly in the psychedelic community, is offered as proof that these individuals have misused psychedelics. This is a circular argument claiming that 1. the psychedelic community does not need to address racism because psychedelics are incompatible with racism, 2. ergo, if one perceives that racism exists, it is because they misused psychedelics; 3. given this, asking psychedelics (or the psychedelic community) to consider issues of racism denigrates the value of psychedelics for true healing.

While the first two arguments in this category imply racism does not exist in the psychedelic community, the third argument implicitly acknowledges that racism is endemic in our community. In this example, individuals warn organizations to "stick to psychedelics" and stop supporting BLM/anti-racism efforts; otherwise, they will lose support and funding from the mainstream psychedelic community. These comments acknowledge that many in the psychedelic community do not value racial equity and would actively fight against racial equity, even if doing so hurt the progression of psychedelic science overall.

MYTH BUSTING

The comments reviewed in this report have contributed to the shaping of eight myths concerning systemic racism, anti-racism efforts, and their relationship to psychedelics and the psychedelic community. For psychedelic science to achieve its full potential, the psychedelic community must deconstruct these myths and understand how they are harmful, misinformed, and racist.

Then we can move forward to become what our community purports to be—racism free—by being an agent of racial justice.

Eight Racist Myths the Psychedelic Community Tells Itself

1. Myth: The psychedelic movement is inclusive and diverse. There is no "diversity problem" in the psychedelic community.
2. Myth: Psychedelics are apolitical. Psychedelic users and organizations should stay out of politics!
3. Myth: Psychedelics unite us as one. Psychedelics elevate us above race, and talking about racism perpetuates the myth that races exist.
4. Myth: Psychedelics will fix society and bring us enlightenment—racial and otherwise. Our best path to combat racial injustice and come together as one is to take psychedelics and let them do their magic.
5. Myth: Comments supporting BLM and anti-racism are just virtue signaling—attempts to look good to the public because these comments are popular.
6. Myth: BLM is a militant organization that uses divisive tactics, and given that psychedelics are the antithesis of divisiveness, the psychedelic community should avoid anything BLM-related.
7. Myth: Psychedelics are illegal for everyone, so the War on Drugs affects us all equally, regardless of race.
8. Myth: If psychedelic organizations continue talking about race and social justice, they will lose mainstream credibility and derail gains already made in mainstreaming and legalizing psychedelics.

We encourage you to make friends with people of color, invite them into your circles, and have conversations about how to bust these myths around racism and BLM. Additional questions to consider are:

1. Who are important POC who have made a difference in the history of the psychedelic movement for you?
2. How can White people in the psychedelic community be better allies in the fight for racial justice?
3. What action can we take to ensure and promote the diversity of the psychedelic movement and the visibility of the many people of color who have played important roles in its formation?
4. What have you experienced in your own psychedelic journey related to aspects of your gender, race, ethnicity, or sexuality?

We must remember: We are all responsible for combating racism in the psychedelic community! We have the ability to create the change we seek and to bring our community to its full potential for equality, justice, and radical healing. At the Chacruna Institute, we are engaged in creating access, equity, and diversity in psychedelic medicine.[5] Will you join our movement?

New Narratives with Psychedelic Medicine

SEAN LAWLOR, MFA WITH MELLODY HAYES, MD

DR. MELLODY HAYES WAS TOLD SHE WOULD NEVER GO TO HARVARD because people from the inner-city high school she attended didn't go there. But Harvard is where Dr. Hayes wanted to go, so Harvard is where she went. After earning a BA in sociology, she continued to medical school at UCSF. Now, Dr. Hayes is an anesthesiologist at Empire Anesthesia, focusing on palliative and end-of-life care.

However, Dr. Hayes' vision of wellness extends far beyond the hospital. In a powerful and impassioned speech at the 2019 Psychedelic Science Summit, held in Austin, TX, Dr. Hayes spoke to psychedelics' potential to help people awaken to their power—a power of love, peace, and joy from which to author new narratives for themselves and for society. She provides a powerful voice for the psychedelic movement that speaks toward and on behalf of all perspectives, including those the movement has yet to adequately represent.[1]

SL: How have you come to see psychedelics as medicine?

MH: I've been reading research for some time about psychedelic use in palliative and end-of-life care, as well as anxiety and depression.[2] When I went through my own experience of depression, I decided to undergo treatment with ketamine. That was a profoundly opening experience. It expanded my understanding of options for treating suffering, opening me to new narratives of understanding emotional distress.

The biomedical model leaves out so much in terms of social, economic, and spiritual distress; therefore, many aren't achieving the relief they need. Understanding our lost connections and social isolation is really important for getting past the underlying causes of our disease.

We're fundamentally social creatures, and our isolation and economic inequality are creating our distress. Part of why psychedelics are such a useful model is that they foster a desire to be in relationship.[3] That is the fundamental medicine.

One thing the scientific community needs to include in discussions on psychedelics is how to bring this healing message to *all* people. History shows it can take 20 years for a novel treatment to reach marginalized communities. How can we include people of color, women, and queer people in the wave of benefit while we still believe there's magic in the medicine?

People call psychedelics *mind-manifesting* medication. What does it mean when everyone has the possibility to manifest happiness, peace, and joy? To experience transcendence, separate from their narratives of suffering?

My treatment with ketamine pushed a mute button on the stories of suffering, and I experienced peace. It's more powerful to create from a place of peace, hope, and joy than from distraction, sadness, and isolation.

SSRIs can numb depression and anxiety, whereas psychedelics can give people that experience of love and peace. With that experience, people can pattern new mental maps, without it, you don't know what to create from.

"How can I choose every day to incline myself more toward peace, love, and joy?" That's how I heal my mind.[4] That's how I create new circuits in my thinking.

If you believe in the law of attraction—"as within, so without"—then, as I become more peace, love, and joy, I attract more peace, love, and joy. As I release aggression, the world becomes less aggressive to me. As I show more generosity, the world is more generous to me.

I trained as an anesthesiologist. I'm passionate about palliative care. But I'm primarily a spirit who believes that matters of love and faith are profoundly important. There will never be a pill or chemical that will give you all you need. We need hugs. We need laughter. We need to be in relationship, in community. We'll never be healthy as we can be when we're isolated. We need societies that are more just and inclusive. That's part of our medicine.

SL: How do we create that?

MH: A lot of things are profoundly societally-determined. Even the number of trees planted in a neighborhood is correlated with health outcomes and levels of violence. So, environment matters.

Let me describe one experimental environment. In the Forced Swim Test for rats, you throw them in water, make it impossible for them not to experience drowning, and they predictably develop a pattern of learned helplessness.[5] You create the experimental circumstances for the outcomes. Others are living in Rat Park, an enriched environment that makes addiction less likely.[6] It's funny that people look at outcomes as unpredictable because these trends are somewhat predictable.

With psychedelics, set and setting are important to produce the experience. We're not all living in the same set and setting. The social violence in this set and setting is producing the psychosis and distress some are experiencing.

SL: You're speaking about experiencing peace and joy, then from that, creating narratives that, through the law of attraction, reflect outward and facilitate healing. In my position of privilege, that feels easy to align with. But I think of someone having this experience and returning to a set and setting of oppression. How can that person avoid getting sucked back into the way life has been?

MH: This perspective frequently comes up: "You wake someone up, then send them back into chaos." Well, you wake up and go back into White supremacy. Does that bother you? Isn't that the same analogy? I would like you to question the lens of privilege of your life, the ease that you're living in. You take it as peace, but is it really as without conflict as you're saying?

There's something fundamentally classist about this idea that treatment with the medicine of love wouldn't work the same in economically-disadvantaged communities. Why wouldn't love and mindfulness work the same? Think of examples like Martin Luther King and Gandhi—no person of color who has become a leader has woken up in peace. They've lived in communities affected by violence, but they've woken up and started doing the work there.

SL: Like Harriet Tubman, as you spoke of in your talk.

MH: Exactly. So why would waking someone up exactly where they're living ever be problematic? Why start with the assumption that an experience of love, peace, and power wouldn't be an effective medicine for all people?

SL: It's not that it wouldn't be an effective medicine, but that more cards are stacked against some people.

MH: That's true. But you wake up into reality and decide to play your hand. If you're going to be a liberal social justice activist, you cannot hold people to the story of oppression as though that is the only outcome that will ever be.

We're in this conversation because we're ready to create new narratives for one another. I don't expect you to behave according to what your ancestors might have done, and I don't expect you to expect me—or communities that look like me—to persist in poverty. While it's useful to understand the steeper climb for people who have existed in oppression, an elevation in consciousness must occur so that we no longer imbibe the stories that predict us to be victims. I rejected that projection, or else it would have become a self-fulfilling prophecy.

Most people look at the world and describe it as it is. Leaders describe the world as it will be. To be a part of social justice work, of shaping new narratives about healthcare, it's important to know where we stand, but more important to keep our eye on what we're going to create.[7] I'm going to create a track—through empowerment, through clinical medicine, through narrative—where people who once felt their only option was disenfranchisement can realize there's a path to empowerment, peace, and joy.

SL: It starts in the heart?

MH: It starts *everywhere*. We have to believe things can change because then we'll act to create change. This "everything-is-messed-up" narrative can overwhelm people, so they palliate instead of curing the system. To really decide you want to empower, you have to believe at the highest level that a fundamentally different outcome is possible.

My vision is that we're going to increase our will for physical safety for Black men and women. I stand for that because if I were to be in my despair about these issues, then I would be lost.

Pain pulls you back to the past. Love, peace, and joy give you a visionary calling that helps you create a new outcome. My knowledge of what we can create is much more empowering than paying attention to the past chaos and violence we've experienced.

There's a Martin Luther King quote that says, "I'm going to stick with love, because hate is too heavy of a burden." If I stay in a place of outrage about where people are in their development, I couldn't engage and say, "That's where you are. How do I get you where you need to be?"

I'm not going to get you where you need to be, because I have to do my own work. But I can encourage you to get yourself where you need to be because I can't keep doing this remedial lesson.

SL: It's not your responsibility.

MH: Right. Your awakening into social responsibility is your job.

I've seen many social justice warriors get sick. They're so in the fight that it takes them out of the game because they're overwhelmed by depression and anger. It can progress into a fixed state of vulnerability of a victim mentality, which is dangerous and isolating.

Although these social realities that marginalized communities live in are real, they're not our only story. If that's the only story you live in, and you don't live in your ability to love and receive love, you'll get sick. But there are ways to be engaged without getting sick in the fight.

Anger is a message about boundaries. It's pure energy. So, it's not the anger we communicate from, but the *wisdom* of the anger. I might think someone else deserves anger, but I don't deserve to *feel* it. I allow the empowerment of my self-knowledge and the wise anger to influence the diligence in which I pursue becoming more bountiful, so that I can hold anger and still contain peace, love, and joy.

SL: Your vision of empowerment isn't the vision of separation and distraction the media promotes. How does one weed through this indoctrinated narrative and stay true to the vision?

MH: You have to focus your attention on what you want to create, not the messages of distraction. White and Black people were telling King that "things will never change." History shows when you're creating what you want to create, people say it's impossible. To them, I say, "I'm not going to pay attention to what you're saying because I see with the clarity of my internal vision. I know you can't see it, but I'm going to walk in this direction, and I'm going to carry you with me."

There's a narrow door and a blade-thin path to the world you want to create. While you're walking the path, people will say, "There's not even a path there. What are you following?" You have to keep walking it, ignoring the *maya* of this world, saying you can't be, do, and have all you're called to be, do, and have.

When I was in high school, my goal was to go to Harvard. Although I was at an inner-city high school in the "hood," I told everyone I was going to go to Harvard. My guidance counselor had no faith in me because she looked where everyone else ended up. I was looking at where I was going to end up. And I got there.

It doesn't matter that some people don't believe in my vision because I decide what tomorrow will be. I'm going to bring people to my vision not because of my outrage but because of my confidence and the abundance we can have.

SL: I saw that in action. Just a few days ago, you received continuous applause through your speech. Your vision is so broad and inclusive that it spoke to everyone.

MH: Thank you. So, the opportunity is to move this from discussion to action. What choices are we going to make?

I applaud you for the choice you've made. I'm going to bless you and say, "May every door be open, that you may amplify the willingness to look at White fragility, the willingness to create a new reality." Because, as you soften, you take responsibility for the shadow, the blind spots to the suffering of others.

There's a great quote I read that privilege means being able to turn away from the suffering of others. When your heart opens, you feel the suffering of others more deeply, a willingness to keep your heart tender and available to a cause higher than yourself.

SL: Do you find psychedelics can help facilitate that?

MH: I find that a spiritual practice is very important. I'm not an advocate for psychedelics per se.

I'm an advocate for waking up and taking responsibility for your life. For some people, that will be through psychedelics. But there's a hazard we should talk about with psychedelics, where people feel good but may not do the work to change their lives and change the world.

The vision is to make a US that's not organized by race, a US where all people have all possibilities of ending up in all destinations. That means our boardrooms look like America. That means our prisons look like America. The stories you have and expect for me are the same as the stories I have and expect for you. We can create that society.

In popular media, there have been many presentations of people of color being exposed to psychedelics. This isn't a niche area anymore. There have been several NPR interviews and several publications in major medical journals. This is contemporary, mainstream medicine.

SL: It's not something you need to tiptoe around in medical communities anymore?

MH: There's a culture within medicine where people don't talk about their suffering. So, talking about my experience with depression is bold of me. However, rates of depression and anxiety are rampant within this profession. My choice to speak publicly about my experience with depression is my choice to be free and bold, to bring these things out of the shadow. It's not going to get better when we keep pretending that we're not in pain.

If we want to shift our discussion about what it means to be human and create more inclusion, people should be able to talk about their anxiety and depression with the same compassion given to someone diagnosed with cancer or congestive heart failure. You don't shame someone for having a heart that's overloaded from too much fluid. But for people with anxiety or depression, whose hearts and nervous systems are literally overloaded—we associate that state with shame.

I've got no shame in my game. I love myself, and I love who I've become by embracing all parts of myself, the shadow, and the light. I share my own story because stories are medicine. Sharing stories is the most important thing we can do to get us from isolation into community, stories about the way life actually is. Everyone's in a Hollywood, Disney version of reality, and that's not what it looks like. Sometimes life is really painful and messy.

SL: And people feel something's wrong with them when their experience doesn't match that narrative.

MH: Exactly. To have people be public and messy and real says, "I'm not alone in this? You have feelings? You want love, too? You hurt, too?" That's healing.

SL: You noted a lack of diversity at the conference. How can we create more inclusion in the psychedelic community?

MH: To create the inclusive community that represents America, every conference has to plan that from the start in terms of attendees, speakers, and everyone in the organization. That means going to colleges and contacting their Black, Latino, and Native American student associations, and giving free tickets for college students to attend.

The vision for clinics would be a sliding scale model, so all people could benefit from these medicines. It's also important these clinics are available in rural areas. Psychedelics are primarily a conversation in urban centers, but rural areas are profoundly affected by depression, isolation, alcoholism, meth addiction, etc.

How do we train therapists from all cultural backgrounds so they can sit and hold space for others? This is a continuation of the need for diversity in medical training at all levels—therapists, psychologists, and physicians. We need people from all backgrounds because your story has a message that heals me, and my story has a message that heals you. We need people who carry all stories inside them, so we witness each other's stories without turning away and instead hold space for one another.

There's a crisis of social connection in America. I don't believe a pill or psychedelic needs to reach everyone, but the message that we need to connect and the impetus to connect does. For people who have lost their way from their knowledge of how to live in authentic connection, an experience with psychedelics could be transformative.

Why Psychedelic Science Should Pay Speakers and Trainers of Color

NICOLE T. BUCHANAN, PHD

DESPITE A COMMITMENT TO CREATING A MORE JUST SOCIETY,[1] MANY psychedelic science organizations do not pay speakers and trainers for their services. I argue that these practices are counter to the values of the field because they have a disparate impact on speakers of color[1] and people with other marginalized social identities (e.g., religious minorities, LGBTQIA+).[2] This is a call for psychedelic science organizations to align their values with their practices by compensating marginalized people for their work as speakers and trainers because failing to do so contributes to social injustices.

THE SOCIAL CONTEXT OF UNPAID LABOR

In the United States (and much of the world), there is a long and painful legacy of coerced work without compensation or benefit. Slavery has a particularly vile legacy in the US, but coerced labor without adequate pay persists, specifically, in the context of prisoner work assignments. These painful legacies should not only be avoided, but actively countered by compensating people for the work they do, particularly when those individuals are members of communities that have historically been enslaved, interned, or had their native lands stolen.

INCOME AND WEALTH DISPARITIES

The legacy of unpaid labor has contributed to income and wealth disparities in the United States. Among those doing the same job, and after controlling for the many factors known to influence income (e.g., education, seniority, etc.), a significant income gap remains between White and Black workers in the US, which is exacerbated further when comparing race and gender groups (a 30% disparity remains in Black women's income compared to

White men's).[3] As a result, the median household income for Black families is $35,400 compared to $59,700 for White households.[4]

Racial disparities in accumulated wealth are even more stark. The legacies of formal (e.g., redlining) and informal policies have exacerbated the wealth gap between Whites and people of color, such that Blacks hold only 7% of the wealth that Whites hold.[5] In 2014, the average median household net worth was only $5,700 for Black households, compared to $130,800 for White households, and the gap has been steadily widening. Even highly educated Black families with two working adults have an average of over $200,000 less wealth than their White counterparts.

The disparities in both income and wealth accumulation imply that not paying individuals for their intellectual and time contributions will have a disproportionately negative impact on people of color because, often, they are already earning less than White peers, and they have fewer wealth reserves to buffer them from the resulting financial strain.

DISPARATE COSTS OF DOING THE WORK

The costs for providing speaking and training services are not equally distributed. People of color and those who hold other marginalized identities are not only more likely to already be earning less than their White counterparts, but Black and Latinx scholars are more likely to be the sole income earners for their family and more likely to be responsible for the financial support of additional family members living outside their homes (e.g., parents, grandparents, siblings). These speakers and trainers are also more likely to have additional jobs to support their families (e.g., clinical work in private practice), which means they experience a cash-flow net loss as a result of traveling and participating in speaking and training events.

Scholars of color are also more likely to be engaging in invisible labor and are less likely to receive recognition for their labor, regardless of its visibility.[6] They are called on to do more service overall, asked to do more service that will not be formally compensated, and explicitly asked to provide free labor in the context of helping their community—which privileged people are almost never pressured to do. Together, this means that marginalized speakers and trainers may be balancing many additional obligations, and these efforts are likely to be invisible to powerful others, underappreciated, and uncompensated.

DISPARITIES ARE COMPOUNDED BY THE ELEVATED PAY OFFERED TO WHITE SPEAKERS

When paid, Whites, particularly White men, are paid significantly more for similar length presentations. For example, they are more likely to have a standard minimum fee required for their participation that is higher than a typical honorarium (a small payment offered as a token of appreciation). The types of activities speakers are asked to do are also impactful. Whites are more likely to be asked to provide an invited or keynote address. These types of speaking requests not only come with honoraria but also provide significant professional benefits because they are more prestigious than being invited to facilitate a training.

It is also important to note that, in psychedelics, many White speakers have revenue streams that are directly benefited when they give talks and facilitate trainings (e.g., selling books) and mitigate the loss of direct payment for their services. Together, the facts that when compensated, White speakers are typically given more money and are invited to do activities that have higher professional benefit, and, when not compensated, they typically have significant revenue streams directly benefited by those activities, makes it all the more important that, as an issue of equity, marginalized scholars are paid for their time and contributions.

IDENTIFYING OUR VALUES

If you want to know what people value, look at how they spend their time, resources, and money. Expecting speakers and trainers to donate their time and expertise without proper compensation communicates that their time, effort, and intellectual contributions are not valued. This message is particularly problematic because people of color are frequently expected to provide such unpaid labor.

Sometimes, a non-cash benefit is offered in lieu of payment (e.g., being allowed to attend the rest of the training, receiving continuing education units (CEU) for their talk, or letting them bring a guest to the fundraiser dinner where they will be speaking). These are unacceptable as compensation if they do not include a direct payment. People cannot pay rent or buy food with CEUs, and even if there are possible longer-term benefits to attending a training, they do not offset immediate expenses accrued by speakers and trainers. The bottom line is this—an honorarium is the minimum that should be offered.

SOCIAL JUSTICE PRAXIS REQUIRES COMPENSATION

Social justice refers to a basic belief that all people are entitled to just distribution of the resources, benefits, and protections within a given society. Social justice praxis moves from the theoretical concept of social justice into the actual behaviors, practices, and policies that facilitate the creation of a just society.[7]

Compensation for services rendered is a minimum requirement for creating a more socially just society. Those committed to enacting social justice to create a more equitable, diverse, and inclusive society must commit to paying marginalized people for services rendered. Ideally, this compensation would be at a meaningful professional rate. At a minimum, compensation should include a modest monetary honorarium.

Psychedelic science has a long history of working to create a just society, such as the efforts of the Multidisciplinary Association for Psychedelic Studies (MAPS) to expand access to psychedelic-assisted psychotherapy.[8] It is time for the field to increase its social justice praxis by recognizing the importance of compensating marginalized people for their work and, at a minimum, committing to giving honoraria when marginalized people generously give their time and expertise.

Considerations for Working with Indigenous People in Psychedelic Spaces & Guidelines for Inclusion of Indigenous People in Psychedelic Science Conferences

BELINDA ERIACHO, MPH

Editor's Note: The following is a combination of two separate pieces by Belinda Eriacho, MPH. In the first section, Bellinda provides guidance for organizations, conference organizers, and sponsors on the inclusion of Indigenous people and communities into psychedelic science events, such as conferences, webinar series, workshops, and other formats. In the second section, she provides guidance to psychedelic therapists working with Native American clients and communities.

CULTURAL HUMILITY

As you create your events, you may be inclined to invite Indigenous people to provide a unique perspective on a specific topic. Partnering with Indigenous communities or individuals takes some cultural humility and competence. This means that the organizers should come with good intentions of honoring Indigenous people's beliefs, customs, and values. It also means acknowledging their differences and accepting them for who they are.

Cultural humility is fundamentally about meeting someone where they are. It requires an ongoing process of self-exploration and self-critique. An openness and willingness to learn from others and listening to understand is key to the success of this relationship. Building a relationship with Indigenous people is not about moving your personal agenda forward. They are not there to fulfill your desire for representation. Rather, it is about working in collaboration for the betterment of all.

Let's cover a few terms to aid in your knowledge of Indigenous people in order to bridge understanding.

TERMINOLOGY

From a Western view, "Indigenous" is synonymous with "aboriginal" and "Native." Indigenous communities, peoples, and nations are those that have a historical continuity with pre-invasion and pre-colonial societies.

In the United States, different terms are used to refer to the Indigenous people that reside here. These terms are "American Indian," "Native American," and "Indian." A further explanation of each of these terms can help organizers properly address these individuals and community in a respectful manner.

The term "American Indian" excludes Native Hawaiians and some Alaskan Native people. "Native American," as a general principle, is a person who has some degree of Indian blood and is recognized as an Indian by a tribe, village, or the United States. There exists no universally accepted rule for establishing a person's identity as an Indian (Native American Rights Fund, n.d.) The term "Indian" has a negative connotation to Indigenous people of North American and is a reminder of the term used by Columbus when he arrived to begin the colonization of the Americas.

The general consensus of members of the Native American tribes in the United States is that they prefer to be referred to by their tribal affiliation, in their native language. For example, Diné or Dineh versus Navajo. As of June 6, 2020, there are 573 federally-recognized tribes in the United States, and each is distinct with their own culture and languages.[1] For many Native American tribes, these names were given to them by those who conquered them or by the US federal government.

WHY IS INCLUSION IMPORTANT?

As stewards of psychedelics and entheogens (sacred plant medicines) for generations, Indigenous people have a unique perspective on plants. This perspective fosters a wider understanding of entheogens beyond the Western scientific perspective. Incorporating traditional Indigenous knowledge into the discussion allows for a wider understanding of the plants and their environment.

Here are some points to keep in mind as you create your event and build a relationship with Indigenous people.

Building a Partnership

- Once a relationship is established, continue to have dialogues with the Indigenous person or community. In order to build trust and strong relationships, consider how you can support their work and projects.
- Have an ongoing dialogue with the invited guest on their perspective on the topic in advance of the event. Come to the discussion to understand through their lens. Here are some questions to keep in mind: What are their trigger points or sensitive topics? What matters to them and their community? What are the benefits for the guest and their community?
- Outside the context of your event, consider how you can offer support at other events and in their communities.
- Provide the opportunity for Indigenous people to have an autonomous and internal discussion among themselves for collective dialogue well in advance of the event.

Building a Relationship

- Take into account that values you consider core to your beliefs, such as "universal rights," gender roles, or others may not resonate with your Indigenous speaker or are not a part of their world dynamics. It is best to get this clarified with them ahead of time.
- Consider honoring your Indigenous guest with a small gift to show your gratitude. In Indigenous communities, individuals can be generous gift-givers. In many of these cultures, it is believed that the richest people are those who are willing to give away or share something from their culture. A small consideration may be flowers with a nice welcome note in their room upon arrival.
- Do our own research about the Indigenous lineage that the speaker is from, and don't expect them to do this for you. Some topics worth researching are: What tribe or Indigenous group are they from? What is the history, and are they matrilineal or patrilineal? What are the issues in the community? What type of government structure is in place to govern their community? What is the role of spiritual, traditional

healers in the community? Does this community use plant medicines and do they have a website affiliated with their work?

- Consider hiring an Indigenous person to support you in doing the research and act as the primary contact with your guest.
- Consider sharing a meal with the Indigenous speaker in a non-formal setting prior to the conference. This allows you both the opportunity to get to know each other.
- Keep in mind that many Native American people may prefer privacy and can be very discrete. So, be mindful of the type of personal questions you might ask. Approach this topic with care.
- Be aware that, although Indigenous people may hold advance degrees, they may choose not to include that as part of their bio. The value of degrees is secondary to building a relationship with others; this demonstrates humility in some Indigenous cultures.

The Event

- As part of your conference logistics, plan on another individual accompanying your guest. This is especially if the invitee is an elder. Consideration should be given to covering the costs of the travel, lodging, and an honorarium. This indicates respect for their knowledge and their willingness to share their knowledge.
- As you develop the schedule for your event, strongly consider placing Indigenous people in the main track of your event or program. In addition, Indigenous people should be considered and included as part of the "scientific" discussions.
- On certain occasions, it might be pertinent to give Indigenous people an autonomous space, separate from the rest of the speakers, aimed at collective internal dialogue among themselves.
- Invite Indigenous people to be an active part of the dialogue in your organization and also ensure that there is a return benefit to them. Tokenization must be avoided. Tokenism is another form of racism. Tokenism allows those in power to maintain the appearance of non-racism

and seem to be champions of diversity, as long as they are beside their recruited Indigenous person.

- As you select an Indigenous presenter, consider whether the individual is representing their individual viewpoint, speaking on behalf of an organization, or speaking as a representative of an Indigenous community. It is especially important, during the introduction of the presenter, that this is communicated to the audience to avoid any misrepresentations or conflicts.
- Be aware that there are individuals who may claim some kind of Indigenous heritage or consider themselves as Indigenous and, yet, they may not be recognized as such by that community. Most Native American individuals in the US have a primary affiliation with one tribe, or what is referred to as "an enrolled tribal member." It is unfortunate there are people who fabricate heritage positions and others who capitalize on Indigeneity as a resource to culturally appropriate. It is critical that the organizers of the event understand the perspectives these individuals are bringing to the event and that these perspectives may not necessarily be the views of the lineage they are claiming.
- Do your proper due diligence when selecting speakers. Be aware there are well-known institutions that may not be organically connected to Indigenous communities who may suggest presenters. Take the extra time to find the potential presenter's position with the institution. Remember that an individual may not necessarily have been given the authority to speak on behalf of a tribal group or a tribal organization. Request a c.v. of their work.
- While arranging accommodations for Indigenous speakers, make an effort to provide the same offering or better as you do for other non-Indigenous speakers. Costs associated with visa processing and travel costs should be fully covered, and you should allow for a per diem while they are traveling to your event. Work the logistics out well in advance of the event. In addition, instead of asking them to catch a taxi, it is better to have a primary contact pick them up and drop them off at their destinations; this is especially true in large urban cities that they may not be familiar with.

- In the early stages of planning, communicate the theme of your conference to the Indigenous speaker and offer some suggested topics that they may speak on. Give them the opportunity to sit with the topic before committing. In subsequent conversations, listen to understand their ideas on the subject they are being asked to present on.
- An academic or scientific psychedelic conference may not be the most ideal event for an Indigenous person to be a part of. You might consider inviting them to provide a workshop where they can share their culture and perspectives or give them the opportunity to do a hands-on event from their culture that may or may not be a part of the event setting. In addition, as a part of the program, an Indigenous person may be asked to do the opening or closing prayer for the event. If they will be doing a prayer or ceremony, ensure that you have the ideal setting for this and offer to get what is needed with their guidance. This may require looking into the site's requirements for the use of candles, smudge, and fire. As an additional consideration, the guest may ask that the prayer not be recorded. Honor their request and work the logistics out in advance of the event.
- When introducing the Indigenous speaker, make sure that you are comfortable introducing them, especially if their bio includes their language. As an alternative, you might do a general introduction and then allow them to make a more formal introduction in their language; you should discuss this with them in advance. To avoid any misunderstandings, it is suggested that you ask them to send you a small bio ahead of time and discuss with them in advance how they would like to be introduced.
- Identify a primary point of contact for the Indigenous speaker throughout the event process to address concerns or questions that may come up. There is nothing worse than being passed from one person to another. This contact should make themselves readily available.
- If the speaker will be sitting on a panel with other speakers, and they are the eldest, allow them to speak first. This is especially true if this person is from the territory in which the conference is being held.

- If the Indigenous speaker is a non-English speaking individual, considerations need to be made for a translator for the primary language of the event.
- It is important to communicate dates, commitments, and deliverables in advance. It is suggested that reminders be sent to the guest.

CONSIDERATIONS FOR PSYCHEDELIC THERAPISTS WHEN WORKING WITH NATIVE AMERICAN PEOPLE AND COMMUNITIES

Native American's World View

Growing up on the Diné (Navajo) reservation, I now realize it provided me with insights that I often overlook as a Native individual. I realized as a young adult in college that the cultural belief system I was raised with was different than my peers. I was raised in a small community where everyone knew each other and where our traditional cultural beliefs were still practiced.

Coming from a health background academically, I realized that in order for providers and practitioners to be effective when working with Native American individuals and communities, they need a basic understanding of the philosophical and social contexts of Native American society. This article is intended to provide insights to non-Native practitioners and others on key points to keep in mind when working with Native American individuals and communities. It is by no means an answer to all situations and circumstances that may be encountered.

Our View of the World

Native Americans as the original people of this country believe that all beings in the universe are sacred: from the tiniest insects that crawl on the Earth to fellow human beings. The elements of air, fire, water, and earth are sacred. Each being is interdependent upon the other for survival and existence. As an example, when I need sage for my Táchééh (Sweat Lodge) ceremony, I do not just pull off to the roadside and pull the sage from the Earth. Instead, I go to a quite remote place in the countryside where the plant grows and when I find it, I offer a prayer and speak to the plant. I tell the plant why I need it and for what I will be using it for. In exchange for its sacred essence, I offer the corn pollen and gratitude. Then, I remove the portion of the plant that I will use taking extra care not to damage the roots of the plant and to ensure that those roots remain intact. This allows the sage to grow more leaves and make it available for the next person.

In the Western world, we separate everything into disciplines and silos —history, medicine, philosophy, spirituality—in most instances the connection between these disciplines is not so obvious or is non-existent.[2] The mystical and sacred experiences of these disciplines are not included. We only view them from our rational, linear perspectives. In the Diné culture, we are taught that all things are connected and what affects one affects the other. For example, if we contaminate our water sources, we will not only become ill, but the animals and plants will also become ill. Another aspect of Diné culture is as human beings we cannot separate our physical, spiritual, mental, emotional bodies, and our environment from one another. My personal experience showed me that to heal from an illness, my physical, spiritual, mental, and emotional bodies had to be in harmony with the environment around me, as well. This healing was done through traditional ceremonies to achieve balance.

In my cultures and in other Native American cultures, we are believers in the unseen powers of the Creator and other sacred deities, and we do not seek to control the ways of the natural world. We believe in the mysteries of life and we allow life to happen without human intervention. In Western society, the belief is that man can control the natural world and create order.[3]

There is an important concept and link between individual's worship and community. In turn, the community carries out forms of worship and ceremonies to aid individuals.[4] As I was going through a series of traditional ceremonies for a severe flare with systemic lupus, I realized that my prayers each morning were to help restore order to my life and with the natural elements from which I fell out of balance. These ceremonies also allowed others in the community to come participate and together we found healing for ourselves and the collective. Each individual's prayers reinforce the bond between human beings and Creator. Prayer makes the individual and community receptive to the blessings that are available if both are responsive to the world around us. These relationship and bonds to each other and to Creator are maintained in a sacred manner.

In seeking happiness in one's life balance, harmony is necessary between the Native individual and the natural elements and forces of nature. In the Diné culture, *hózhó* is the holistic environment and a representation of beauty, harmony, and well-being. Our beliefs are to take care, understand, and learn to be a part of the ecological relationship with everything around us. By doing this, we learn to survive in the natural world. Without it, we are alone and do not have the support of the universe.

Typically, when a non-Native individual meets a Native individual, their impression is that the Native person is very reserved and stoic, similar to how it is portrayed in the media. Native people enjoy humor and humor is an integral part of our lives. Humor is at the heart of our resilience. Humor is used as a way to teach valuable moral and social lessons. In addition, humor can be used as a way of healing from personal or historical trauma. Humor can also be used to foster relationships between non-Native and Native people. In many of our cultures, there are deities who are the personification of a clown and serve to remind us not to take ourselves too seriously.

SUGGESTIONS FOR WORKING WITH NATIVE AMERICAN INDIVIDUALS AND COMMUNITIES

There are some points for therapists and practitioners to keep in mind when working with Native American clients and communities. Below are some considerations:

- Trust and respect are the keys to building a strong rapport. It is important to understand throughout our history with non-Native people that our relationships were typically fraught with deceit and mistrust. As a result, our natural tendency and response to someone offering to help us is skepticism. Therapists and practitioners will need to exercise patience and over time for the relationship to flourish.

- Native American individuals are private by nature and do not typically disclose their personal lives. Growing up in a small community on the reservation where everyone knows each other and should not be taken too lightly. There is often reluctance for a Native individual living in a small community to share their personal lives openly, especially with a stranger. There may be fear from the Native client's perspective that others in the community will know their business. Therefore, it is important that precautionary measures are taken to ensure each individual's privacy outside of the clinical and therapeutic settings is maintained.

- Therapeutic models that allow for Native American traditional practitioners to be a part of treatment plan processes have proven to be beneficial.[5] In addition, allowing Native clients to use health practices (such as sweat lodge or talking circles) can improve the treatment outcomes. The concepts of mental illness and associated disorders have a different causation and remedies in Native American cultures, and this understanding should be included in the treatment plan as much as possible.[6]
- Spirituality is not separated from our physical, emotional, and mental bodies. Hence, the treatment plan should be culturally specific.[7]
- Educating ourselves about the Native American culture(s) for which we will be providing care is fundamental to establishing rapport with clients and community. There are 573 federally recognized tribes in the United States each with its own culture, traditions, and languages.[8]

Native American individuals and communities are unique in their beliefs, cultural traditions, and languages. However, there are common belief systems shared by most Native American cultures. Our view of the world is based upon being a part of a larger microcosm of the natural world. The interconnectedness of our existence is based upon the mutual respect of all beings and embracing the mysticism of life. This interconnectedness includes the relationship each individual member and its community and this reinforces the bond between human beings and Creator. The goal in life from a Diné perspective is walking through life in hozho, living in balance, and harmony. Furthermore, humor is a part of our resilience and moving toward hozho.

As therapists and practitioners, there are fundamental aspects of the Native American culture that one should be aware of when engaging in relationship with Native Peoples. This includes trust, respect, confidentiality outside of the context therapeutic setting. Therapy should also be culturally based and inclusive of Native American traditions and culture because it is important to gain an understanding of the culture in which one will be interacting.

SECTION TWO

Perspectives on Cultural Appropriation, Colonialism, and Globalization of Plant Medicines

My Degree Is From the Forest

LEOPARDO YAWA BANE

(Transcribed, translated, and edited by Chris Dodds)

I BELIEVE THAT THE ISSUE OF AYAHUASCA IS, IN ESSENCE, AN ISSUE OF human rights, as humans of this Earth. I believe that we all have the right to seek healing from the natural plants that we share this planet with, and as Indians and stewards of our lands, and guardians of our medicinal plants, I believe that it is our right to be able to travel and administer these cures without fear of persecution.

While the global spread of Ayahuasca and Indigenous culture is something that I support and has given me great joy and opportunity, there are some things that worry me. My main concern is that our people and our culture will be forgotten as the White man gets carried away with being an "authority" on Ayahuasca. "Grandmother," as some call it, is not meant to be yet another medicine that gets plucked from the forest to enrich pharmaceutical companies and leave Indians, the true guardians of this sacred medicine, behind. Most of you do not have this intent and may even think that is an exaggeration, but our people, whether in North, Central, or South America, have been lied to ever since the White man landed on our shores some 500 years ago.

PSYCHEDELIC SCIENCE 2017 CONFERENCE

I recently attended and presented my story and beliefs at the Psychedelic Science Conference, presented by MAPS and the Beckley Foundation, in Oakland, California. I think that the idea of the conference is a beautiful thing—in that it's basic focus is how we can help people with the use of psychedelic substances and, as a "curandero," that is exactly what I do. But that is not to say that I don't think there was a noticeable lack of Indigenous representation.

Of the entire conference, with more than 60 speakers from around the world attracting over 2500 attendees, and roughly half of the panels on the Plant Medicine Track being about ayahuasca—I was one of the few Indigenous speakers; and, to be perfectly honest, I was upset over this. I know I am not the only one—a woman who identified herself as an Indigenous person even brought it up during one of the other panels and that was even mentioned in The New York Times coverage of the event. I was not invited to be a part of any of the main panels, and rather was delegated to the smallest stage as part of the free public marketplace put on by the Psymposia (to whom I am very grateful).

I hold no grudge, and it is not my wish to only complain or be ungrateful, yet, to quote the Indigenous woman who spoke up, "What I want to say is that cultural appropriation is a form of re-traumatization to Indigenous people." For 500 years we have been traumatized and now—thanks to ayahuasca—we are finally gaining attention and respect, but to be largely ignored at such a prestigious and monumental conference is highly traumatic.

After the presentation on Sunday of the conference I was happy to see that my message resonated with so many people. One young attendee said to me, "This is exactly what I've been looking for all weekend! Thank you, thank you, thank you for coming and sharing with us."

Another said, "I am so happy to have heard your voice on this. I kept wondering why there wasn't any Indigenous representation at any of the main panels."

I AM NOT A SCIENTIST; I AM AN INDIAN

I am very thankful for the opportunity to have presented and connected with these people because, up until that point, I started to think that perhaps there was no longer a need for the Indigenous pajé (shaman). What I felt at the conference was that was that we weren't needed, that all of these doctors, neuroscientists, chemists, and academics had apparently figured it out and moved on without us. Luckily, I said this to some of my txais, and they reassured me that I was wrong and said that "Now, more than ever, we need the pajés to help navigate these strange times."

Perhaps, for some people, they need to have a previously unknown substance like Ayahuasca validated by those with credentials they understand like a PhD. Yet, for an Indigenous person, it is almost impossible to attain such an advanced degree, especially if the majority of your life is spent in rural and isolated areas, like the Amazon, where our communities are located. For most of us, the forest is our university—and in that case, we have many advanced degrees.

THE LIFE OF A SHAMAN

Recently, through my work, I have realized that there seems to be some sort of illusion of "rockstar shamans" and that all we do is travel the world and are overpaid. That is akin to saying professional athletes who come from impoverished backgrounds such as favelas, ghettos, or government housing projects have everything so easy. This grossly ignores all the work they had to do and all the racism they had to endure to get to where they are—even Lebron James, the most famous and well-paid basketball player in the world, still deals with racism.

Yet, in almost all societies, the Indigenous are regarded as the lowest of the social classes. Outside of a few countries, the work we do is considered illegal, and the pay we receive is very little when considering that many of us have large families, even entire tribes, to support. The travel and workload can be never-ending and we often spend long stretches of time away from our homes and families. On top of that, our people and native homes are under constant threat.

TXAIS AROUND THE WORLD

I know that not all White people, or all Westerners, are bad—far from it! In my language we have the word "txai" (pronounced "chai") that means half of me, is half of you, and that half of you is half of me. During my travels, I have made many, many new txais across the world. It does not matter where you are from, or what color your skin, it is who you are in your heart that matters. I implore you to be a txai by being aware of our history and culture and supportive of our struggle for we are all on this planet together ("estamos juntos," as we say in Portuguese). If we wish to thrive, not merely survive, in the 21st century we must join forces and build bridges, not walls, between our communities.

We have much to learn from each other, and the first step is to recognize that this is a two-way street. There is no more time for the "I am right, and you are wrong, mentality." It is clear to see that there is much pain and suffering in this world and the only way to overcome our problems is to work together. I am a father to three children and my greatest wish for this world is that all the people can come together to create a brighter future.

Colonial Shadows in the Psychedelic Renaissance

DIANA NEGRÍN, PHD

IN APRIL 2020, I JOINED THE TWO-DAY PSYCHEDELIC LIBERTY SUMMIT, where the voices of several Indigenous participants from Colombia, Brazil, and several tribal nations in the United States discussed their concerns over the parallel trends in decriminalization efforts and the expansion of the use of sacred plant medicines. These medicines and the cultural practices that have sustained their safe and sustainable use are now, more than ever, being consumed by a global public. Many Indigenous peoples argue that these plants and their spiritual practices are being appropriated. At the same time, their native territories continue to be encroached upon for other global consumption items like minerals, fuel, and beef.

Why these strong words from Indigenous peoples? Much of the psychedelic public shudders at the idea of having their practices labeled exploitative, unsustainable, or unethical. Many are asking why racial and ethnic boundaries are being placed around legitimate versus illegitimate uses of what are called sacred plants, ancestral medicines, or entheogens. Before I continue, I want to note that there exist a great variety of non-Indigenous communities that interact with these medicines outside of the more visible platforms I discuss here. The point of the present critique is to consider the ecological and social vulnerabilities that are ignited by the current surge of interest in these plants as they are marketed and inserted into distinct cultural settings.

The reality both for Indigenous observers and many dedicated to environmental and human rights advocacy is that the global consumption of these plants reflects relations of power rooted in the racial order created with colonialism and the ongoing destruction of the very habitats these plants are endemic to. In other words, seemingly benign practices aimed at therapy or recreation are endangering the very plants they are celebrating.

THE COMMODIFICATION OF THE SACRED AND THE RUIN OF THE LAND

"Your authentic peyote ceremony experience will involve true desert camping." San Miguel de Allende to Real de Catorce, round trip. Price not listed. Bring a sun hat. Tulum, Cuenca, Iquitos, the jungle of Costa Rica, Barcelona, the desert of New Mexico. These are all places that you can visit to experience the potentially healing powers of sacred medicines. For $1200 dollars you can enjoy a two-day "Don Juan Teachings" peyote ceremony organized by a Polish man with the help of a "Huichol shaman" in the comfort of a hotel in Cancún. If you are looking for a deal, elsewhere along the Mayan Riviera, in Tulum, a young woman who self-identifies as a shaman provides peyote ceremonies accompanied by "Wixárika chants," for only 45 euros! Or head further south for any number of retreats where you can rotate ayahuasca, San Pedro, peyote, kambô, and rapé. Choose one plant or many—it is a buffet of interchangeable ecologies and cultures administered through the hands of people who provide biographical abstracts to assure their capacity to safely officiate ceremonies with these plant and animal medicines.

From platforms like Instagram to alternative retreat sites, these plants and the cultural practices that promote their use are on full display. The women are sexy, there are sweat lodges and tipis, chanting, eco-villages, yoga, massages, and vegan food.

The cliché remains true: We live in a world in need of deep psychological healing, and many of us live in societies and within cultures that do not offer us the answers to our crises. As women, we continue to be oppressed and desire liberation and spiritual connection; as urban citizens, we desire health and a landscape beyond asphalt. It sounds like a repeat of the 1960s and the first boom in psychedelics. Fifty years later, the existential search continues, and Indigenous cultural practices once more are understood and consumed as alternative ways of being in the world and with the planet. The second boom is here, but I fear that it is a boom on steroids with corporate financing, patents, and bioengineering on the horizon and a much more mobile (at least pre-COVID) community of seekers.

With this in mind, what does it mean to partake in the buying and selling of these sacred plants?

What are the implications of moving these "ancestral medicines" across various geographies? And how is it that those involved in these activities utilize notions of indigeneity, and often Indigenous associates, to add value to their circles and retreats, irrespective of authenticity, conflict, and power?

As a student of the history of Wixárika territory and culture, I have closely followed efforts to protect the sacred pilgrimage place of Wirikuta from the devastation of mining and agro-industry.[1] It is this land in the Chihuahuan Desert that peyote is endemic to. For much of the consumer public, these medicines are divorced from their habitats, and this very fact is dangerous to their survival. In the case of peyote, I have yet to see its scarcity or slow growth cycle be properly acknowledged. Peyote's condition is abstracted through a language of abundance, for it is abundance that the seeker desires. On the other hand, for most Indigenous communities, there is a clear relationship between these sacred plants and the historical and spiritual geographies they are rooted in. Territory and culture are not so easily disentangled.

Since 2010, the Natural Protected Area of Wirikuta and its surrounding region in the plateaus of north-central Mexico has faced the threat of a landscape that is changing through macro practices, like mining—the harms of which are heightened through the effects of climate change—and through micro-practices, as seen in the over-extraction of peyote. Scale is undoubtedly important: Pipelines and industrial egg farms pose an immediate damaging effect to the land, while the effects of mining take a little longer to manifest but remain in the water and soil for generations. In this way, peyote extraction by "hippies" is less pressing; yet, some researchers are beginning to demonstrate that extraction of peyote is also being fueled by global consumption of the plant in powder form, which allows it to travel greater distances, making it into the parties of Ibiza and a variety of ceremonies and retreats throughout the Americas.

As Wixárika lawyer, Santos Rentería, emphasized at the Chacruna's Sacred Plants in the Americas Conference in 2018, peyote is part of "a whole"; it is not merely a plant, it is a being that is part of an ecological and cultural landscape dating back millennia.[2] All elements affecting the individual peyote are also transforming the landscape as a whole. In the end, both the landscape and the individual plant are under threat, and their conservation must attend to this symbiotic factor. The struggle to protect the Chihuahuan Desert's biodiversity is deeply related to the long-term conservation of peyote in its endemic geography, itself, the richest source of spiritual knowledge.

THE IMPERATIVE FOR HORIZONTALITY AND DECOLONIZATION

As Eduardo Galeano wrote nearly 50 years ago, across all Indigenous territories of the Americas run similar veins that have lived under assault for several hundred years. Yet, to this day, we often do not mention that the violence perpetrated by colonial and capitalist structures of economic and political power is sometimes accompanied by another side, represented by awe, curiosity, and even desire for the "otherness" of Indigenous cultures. This other side can be observed in the amply documented (mis)uses of sacred Indigenous plants and traditions, and it is past time that the global public interested in these matters makes space, listens to, and respects the perspectives and autonomy of Indigenous communities.

I return to the closing remarks at the Sacred Plants in the Americas Conference when Aukwe Mijarez of the *Consejo Regional Wixárika* (Wixárika Regional Council), moved the audience to tears when she reminded us that, as a Wixárika person, to not find peyote is to face the deepest sadness, the inability to *cumplir,* to meet agreements with the ancestors and provide the foundations for collective livelihoods: "More than just a plant, this is about the survival of a people! We do not want to be dressed up to be an exotic people, and for our culture to be prostituted and sold, but to be defended!"[3]

WHAT DOES A WIDER NETWORK OF CONSUMERS MEAN IN THE FACE OF SCARCITY?

I would like to suggest that the broadening global interest in these medicines requires a moment of pause and education in order to put into perspective the ecological and cultural footprint that accompanies the consumption of sacred plant medicines. Many organizations and individuals have spent decades researching, advocating, and caring for these plants through various cultural perspectives and from various geographical locations. Increasingly, the "biocultural" approach promises a model for conservation and ecological restoration projects that are premised on multiple complementary knowledges and a greater balance in the relationship between humans and ecologies. It means defending the lands that sacred plant medicines are part of while respecting the knowledge of Indigenous peoples who offer generations of investment in understanding these medicines.

Undoubtedly, there is a deep need for psychological and physical healing for a much larger group of people who do not currently have access to the potentials offered by some of these medicines. This, too, raises questions about equity. Luckily, this matter is being addressed not only by a network of people who have long understood the benefits of these medicines for psychological and physical healing within Indigenous communities but also by practitioners of color who are committed to underserved populations. As we move forward with any conversations on decriminalization and conservation, these voices need to be front and center.

Across our differences, I see hope if we are able to agree on some of the following premises:

1. To protect the habitats that are the homes of these sacred plants and animals, and through social and political organizing, bring together concerted efforts toward the conservation and restoration of these very territories.
2. Create and provide cultural and environmental education for the broader public to understand and respect mandates around the specificity of each medicine, its history, and the cultural context that frames its study, conservation, and use.
3. Prioritize practices that come from a place of humility, collaboration, and horizontality.

Those of us involved in these discussions and practices must work to better understand that coloniality is premised on the ongoing displacement, appropriation, and distortion of Indigenous land and culture. Following this, decolonization can only begin by listening and carefully working toward collaborative models across geographies and identities, always with full recognition of the autonomy and leadership exercised by Indigenous communities who have been the greatest custodians of these medicines.

Cultural Appropriation & Misuse of Ancestral Yagé Medicine

UMIYAC

(Translated by Leonardo Hum. Edited and revised by Riccardo Vitale.)

SPIRITUALITY AND YAGÉ

Territory, culture, self-government, and spirituality: that's what's most important for us.

The *curacas*, our men of wisdom or traditional doctors, harmonize relations within the community and heal illnesses. The *curacas* know the territory, know and communicate with the spirits of Mother Nature, and feel the sacred energy sites of the territory. The curacas have studied and practiced the medicine for ten, twenty, thirty—even forty and fifty years. Often, they come from lineages of medicine men and medicine women. The curacas are our spiritual neurologists. Learning the medicine is an arduous lifetime mission.

Formerly, the curacas were "tiger-men": powerful medicine men capable of transforming into tigers, birds, or reptiles. Today, because of environmental contamination, loss of knowledge, religions, war, plundering, and technology, these skills and other knowledge are in the process of extinction.

When there is harmony in the territories, the inhabitants of the community participate in the yagé ceremonies, contribute to the *mingas* (cooperative and voluntary work for the common good), respect the culture, and do not succumb to vices, drunkenness, family violence, and the extreme poverty of those who spend all of their income drinking excessively. The leaders of the *cabildos* who take yagé under the supervision of the curacas are less susceptible to the corruption that is prevalent in local government.[1]

Yagé medicine is our process of collective and individual spiritual healing. The ceremonies are sacred and social. Friendships are built, we laugh, analyze the problems, and find the solutions. We share knowledge and experiences. In the long nights of rituals, the fire and the sacred and powerful yagé plants of our forefathers illuminate our thoughts and clean our vision. There is also a lot of suffering. Ancestral medicine is a process, with each person in his or her internal and sacred path, facing the difficult trials that yagé presents.

Only the elders, the curacas, the true holders of knowledge, can recognize illnesses and heal those possessed by malignant spirits. The others, the younger drinkers of yagé, are followers. We drink to improve as persons and spiritually. The medicine and the curacas help us leave behind our bad habits. We cry, we suffer, and we get up again, stronger. *Indios* live thus, learning to suffer, overcoming the trials life gives us. In order to learn and improve as a person, it is necessary to drink for many years, to continue drinking, and to grow spiritually, one little step at a time. This is the endless path of learning. The route of yagé is difficult and requires much personal care; it is essential to respect the norms and instructions given by the elders, the curacas.

CULTURAL APPROPRIATION AND MONETIZATION OF ANCESTRAL KNOWLEDGE

> UMIYAC does not recognize the legitimacy of persons foreign to the communities (neo-shamans or charlatans), or Indigenous drinkers with limited experience, as practitioners of the medicine. A preoccupation for UMIYAC is the proliferation of neo-shamanic practices, the commercialization of the sacred plants, and the processes of cultural appropriation and of extraction of local knowledge. This includes academic research, when it is not clearly aimed at strengthening the Indigenous organizations' autonomy and self-government.

People come, settlers, strangers, Whites and Indios. Some come from the cities. We always receive them as friends because the doors of healing are open to those who need it. They come to the territory, to the curaca grandfathers, and put on feathers, sacred crowns, necklaces, rattling collars, and they look like tiger-men, *taitas*.[2] The grandfather watches and remains silent. Then the medicine is distributed, and the grandfather says that if the newcomers are crowned taitas, then we must drink as crowned *taitas,* and he fills the cup to the brim. The grandfather, you see, has a naughty sense of humor. Then the suffering begins. The charlatans twist in pain on the floor—they cry, they vomit, they soil their pants. Some ask the forgiveness of God. In the morning, very early, with nothing to say, they leave in silence. In ancestral medicine, it is necessary to learn respect towards the curaca elders, respect for yagé, and respect for oneself.

One has to be humble before yagé.

Blind ambition for business and money foments bad practices and feeds the commerce of sacred medicine. There are members of the communities and also "Whites" who go about Colombia and the world wearing crowns and doing commerce in yagé—some even call themselves taitas.[3] They are people who do not have the knowledge and are not authorized by the true taitas or curacas to practice the medicine. It is sad, for this constitutes an appropriation of our ancestral practices, and it is lack of respect for our cultures, for sacred medicine, and for the curaca elders. Here, there are drinkers who are inhabitants of the cabildo who have been drinking yagé and practicing the medicine for years and years, whom we do not call taitas and who, out of humbleness, do not care to be referred to as such.

Bad practices of yagé medicine offend the curaca grandfathers and cause serious problems such as rape, sexual assault, and even deaths. This harms the image of our communities and causes suffering and more illness in those seeking healing.[4]

THE TERRITORY

Our task as practitioners of the medicine and members of Unión de Médicos Indígenas Yageceros de la Amazonia Colombiana is to safeguard and strengthen the territory, to strengthen the culture and our self-government, and to strengthen the ancestral medicine we inherited from our forefathers that remains heritage of the Indigenous peoples—and all of humanity.[5] Every time a hectare of territory falls in the hands of mining, logging, or oil companies, land grabbers, or drug traffickers, the Indigenous communities and the rest of humanity lose ancestral knowledge that is essential for the survival of the entire planet.

Notwithstanding national laws and international treaties that safeguard our cultures and territories, there are many pressures that put at risk the Amazonian life systems.[6] The name of this plague scourging our territories is "extractive economy." This is the social relationship that has historically linked our cultures with the "outside world." It is the pillaging of resources, fauna, flora, minerals, hydrocarbons, people, and knowledge, as in the case of the pharmaceutical industry and even extractive anthropology.[7]

SELF-GOVERNMENT AND GRASSROOTS ORGANIZATIONS

The peace agreement between the government and the FARC guerrillas has meant an important step for our communities, historically victims of the armed conflict. It is also certain that now, new powers and economic interests, such as legal and illegal mining, drug trafficking, and the hydrocarbons and livestock industries, among others, are advancing in taking control of territory with the aim of filling the gap left by the insurgency.

Hence, the importance of the process of strengthening grassroots organizations, Indigenous self-government, the cabildos, rural associations and networks, and organizations of spiritual authorities, such as Unión de Médicos Indígenas Yageceros de la Amazonía Colombiana (UMIYAC).[8]

Self-government is the expression of the natural laws of Indigenous communities. Through the construction of self-government and the strengthening of grassroots organizations, the Indigenous communities fulfill their role of safeguarding the territories. This process is the response of the Indigenous communities to the call of Mother Earth for reestablishing ancestral, spiritual equilibrium and a response to a "post-conflict" scenario, under which the assassinations of human rights and environmental defenders continue to increase.[9]

Mazatec Perspectives on the Globalization of Psilocybin Mushrooms

ROSALÍA ACOSTA LÓPEZ, INTI GARCÍA FLORES,
SARAI PIÑA ALCÁNTARA

THIS ARTICLE IS THE RESULT OF CONVERSATIONS NOT ONLY BETWEEN ITS authors but also with Mazatec friends in the region who have at different times expressed their concern about the "little ones who sprout from the Earth," especially in the face of an emergent psychedelic capitalism. To them, we give our respects.

In the Sierra Mazateca of Oaxaca, health problems are treated by mushroom healers, suction doctors, body cleansers, and snake-oil shamans who remove illness from people using various medicinal preparations such as mushrooms, morning glory seeds, shepherdess sage, cow's tongue, rabbit yam, deaf man's yam, hangover mint, and so on. In this study, we will focus on the so-called "magic mushrooms," known as *Ndí Xijtho* or "little things that sprout from the ground" in the Mazatec language. In order to appreciate this people's connection with the sacred mushroom over the past centuries, one must understand the relationship that various Mazatec families maintain with Ndi Xijtho as a means of healing and divination.

The Ndi Xijtho help the Mazatec communicate with the Creator-being who gave life to humanity; the same being who cures them from evil and pain. Ndi Xijtho are sacred to the Mazatec because they represent the strongest expression of the Mazatec spirit. The mushrooms are always used in pairs in Mazatec ceremonies, representing the duality and fluidity between the masculine and the feminine, between father and mother. In turn, Ndi Xijtho is the body of God and allows the deity to take possession of the body of those who ingest it: God's blood flows in their veins, God's saliva makes them speak aloud during the ceremony; the sacred mushroom are the means by which they speak with God and other beings of the sacred world, *Són'nde*. It is the Ndi Xijtho who show them the right path and who give them good luck. They are not consumed for pleasure or fancy, but because there is a need to heal. The mushrooms are medicine for the Mazatecs, and this ritual is so sacred that it can only be carried out on certain days of the week.

Regarding the ritual use of the sacred mushroom, the ceremonies follow the agricultural, religious, and festival calendar of the Mazatecs. In the twenty-day month of Chan-Majti ("Angry Month"), it begins to rain and thunder, hastening the birth of mushrooms. The Mazatec say that thunder makes diamonds fall from the sky (*Chosinle-Ngami*). Thunder is caused when Chikon-Tokoxo, the supreme supernatural being the Mazatecs (from Huautla), pay tribute to, slashes lightning bolts with his axe. That is why it is said (*Jacha Lechaon*) that during this twenty-day period, you cannot take offerings or prayers to the sacred hill of Chikon. During this time, people begin preparing medicinal tobacco (*Jna-Jno*), referred to regionally in the Sierra Mazateca as *piciete*. Tobacco leaf is ground and mixed with garlic and lime and carried in small packages for the protection of the owner, especially during ceremonies. In the twenty-day month that follows, Chan-Sinda ("Month of Toil"), the spring of the sacred mushrooms (Ndí-Sijtho) begins. Mazatec elders say that mushrooms are born miraculously, that they are a gift from the gods, sent from the heavenly domain by thunder (Naí-Chaon).

When the Mazatecs perform healing rituals, a "man or woman of knowledge" (Chjota Chjine or Chjon Chjine) gathers the sacred mushrooms where they sprout in places that only they know by means of a special ritual seeking permission from the mushrooms' owner (Chikon). The mushrooms can only be picked from the ground by children, whose purity and virginity does not corrupt this sacred harvest, ensuring truly beneficial rituals and cures. The mushrooms were traditionally collected at dawn on a full moon and carried with great care, trying to avoid any inauspicious omens along the way (encountering a dead animal, passing by a house where a wake is being held, seeing an injured or sick person, meeting a pregnant woman). These precautions ensured that the sacred mushrooms were not contaminated or spoiled before being delivered to the Altar of Knowledge, which is referred to as a "clean" or "transparent" ritual table (*Yaa mixatse*).

The ceremony traditionally begins at night and ends at dawn. Ceremonies are always presided over by a person of knowledge (Chojta Chjine) who is responsible for purifying the mushrooms with copal incense and using medicinal tobacco to anoint the person who has come to the ceremony to be healed or repair some problem in their life. The Chjine "marries" the mushrooms into pairs (father and mother) in ritual language and gives the patient their dose. The Chjine also eats his or her portion of sacred mushrooms during the ceremony, singing, praying, speaking with the Chikones, and fighting with evil spirits. Ceremonies are carried out in the Mazatec language, with some parts in Spanish. At the proper time, the Chjine intercedes with the gods or supernatural forces to cure the patient if they can be cured. The Chjine speaks intensely, but custom has it that it is the mushroom who is actually speaking, leading the ritual participants where they want them to go. At dawn, the mushrooms finish their work. Before and after the ceremony, four days of sexual abstinence and a special diet are kept. The Chjine recognized by Mazatec communities have preserved this collective and ritual knowledge, despite "modernity," and have resisted misrepresentations by certain tricksters. For this reason, they call on foreigners, as well as their own community members, to show respect for profound wisdom, rather than vandalize customs.

FIRST ENCOUNTER BETWEEN WESTERNERS AND NDI XIJTHO

In 1957, New York banker and mycologist Gordon Wasson published a now-famous photo essay in *Life* magazine, making the Nndi Xitho ritual known to the general public far beyond the handful of mycological specialists who had been aware of the practice previously. He described an evening he spent in 1955 with his wife Valentina Pavlova and Chjon Chjine María Sabina in Huautla de Jiménez. The essay describes a ritual that had previously been thought extinct, but was now revealed to have been jealously guarded, since the mushroom belonged to the sacred realm.

Wasson was later criticized by scholars for his lack of ethics in revealing the "little ones who sprout from the ground" and publicizing the image and songs of María Sabina without her full consent. However, given the increasingly globalized world, sooner or later, with or without Wasson, these sacred rituals would have been discovered. With this international publicity, foreigners with diverse interests—whether for research, curiosity, or spiritual or mystical needs, began coming to the mountains of Oaxaca seeking contact with the little ones. In this regard,

Wasson himself noted: "These words make me tremble: I, Gordon Wasson, am responsible for ending the religious practices of Mesoamerica that go back thousands of years.... I never doubted what I should do. The sacred mushrooms and religious sentiments that they embody in the southern highlands of Mexico were revealed to the world, just as they deserve, no matter what personal price I had to pay for this." Wasson provides a clear example of the paradox facing scientists from various disciplines, and even Indigenous intellectuals: How much of such investigations should be revealed? Are all studies validated in the name of science? Is the agency of these communities and groups taken into account with respect to their cultural practices?

The arrival of foreigners took place in a complex historical context because, alongside the arrival of the counterculture movement in the 1960s, increasing numbers of roads and highways were being built in the Sierra Mazateca at the time. These opened a remote, mountainous region evermore to capitalist logic and government initiatives, further facilitating the arrival of *güeros*, as tourists are called. Over the years, a special kind of tourism began to develop, catering to those who sought to have contact with the "little ones" and the Chjon Chjine or Chjota Chjine, inspired by the magazine story about María Sabina. She became a key figure for outsiders, but was ostracized and reproached by her community for revealing the secret of the mushrooms.

This tourism trade resulted in sacred knowledge (the mushrooms and the ceremony) being offered to outsiders, initially, under a logic of reciprocity, since early exchanges were based on barter, but that soon became monetized. This commodification of sacred heritage resulted in the rise of autonomous neoshamans who worked for tourists, and sacred mushrooms gaining a monetary value for foreign tourists as well as locals. Those who offer the ceremonies were viewed ambivalently, either valued or despised by different segments of the Mazatec community. The situation was further complicated when Huautla was included in a government program of "Magical Towns," as the Mexican state attempted to appropriate local cultures to promote tourism.

EFFECTS OF THE PSYCHEDELIC RENAISSANCE: NEGATIVE BIOCULTURAL APPROPRIATION

Western interest in the use of Indigenous peoples' sacred plants has various motivations, from spiritual and mystical to recreational. In recent years, as restrictions on studying many substances have been relaxed and worldwide psychedelic research has entered a kind of renaissance, interest in psilocybin has increased. Organizations like Compass and Usona have contributed to growing interest around the use of the substance for spiritual, medicinal, and clinical purposes. Yet, it is worth asking: Is this interest only for therapeutic uses? Are we facing a more generalized opening of a psychedelic market? How can Indigenous peoples' knowledge about psychoactive plants be taken into account without falling into negative biocultural appropriation?

Over the last two years, there has been a growing influx of foreigners to the Sierra Mazateca region seeking to learn about the ritual of the Ndi Xitho and gather spores of the mushroom (*Psilocybe caurelescens mazatecorum*) to take back to their countries of origin. Several claim they want to help other people in the West to cure their ills or, as Wasson once argued, "help the ritual survive." But hasn't the ritual survived for more than 500 years, since the beginning of the colonial process through today? Still others argue that mushrooms belong to all humanity. In the case of *Psilocybe* this is partly true, since there are various species worldwide, although many came to be in different countries through acts of biopiracy. However, in the face of this expansion of the use of mushrooms, how can we exercise a more horizontal learning process without falling into unequal power relations and appropriation? We do not intend to say that non-Mazatecs should be prohibited from experience with the "little ones." Rather, we want to make it clear that various kinds of experiences can be valid, especially those that are carried out with adequate respect and without the desire to profit from the the *Psilocybe* mushrooms of the region.

The Mazatec people have shared their construction of knowledge around the Ndi Xitho for years. We believe that, ideally, Western researchers who have studied psilocybin for different reasons would commit themselves to creating forums for dialogue, encounter, and discussion with those who have been working for generations developing an entire therapeutic tradition with the Ndi Xitho. These exchanges should move beyond the folkloristic attitude towards the ritual, appreciating the validity of the holistic therapeutic component of the ceremonies, which are intimately related to ways of being and sensing in the world. Finally, these approaches should not focus on only a few specialists in the region, which would reproduce unequal power relations while ignoring the important fact that this sacred legacy is, above all, a collective heritage.

SECTION THREE

Psychedelics and Western Culture

The Revolution Will Not Be Psychologized: Psychedelics' Potential for Systemic Change

BILL BRENNAN, PHD

The oceanlike immensity of joy
Arising when all beings will be freed.
Will this not be enough? Will this not satisfy?
The wish for my own freedom, what is that to me?
—Shantideva[1]

HOW MANY OF US PSYCHEDELIC USERS CAN FIND SOME EMPATHY FOR THE idea of "putting it in the water supply?" Not for the reprehensible tactic, but for the underlying fantasy that, if enough people were initiated into the mysteries, we could save the world together. When we take a psychedelic, we feel this possibility deep in our bones. We sense the lock-and-key fit between the immediacy of our psychedelic experiences and the urgency of the species-level problems we face. These substances change us, while showing us the need for a global shift that reaches far beyond our individual selves. Our own healing leaves us with a more impassioned ache for the systemic changes that would heal our collective life as well.

For many, our collective liberation will come on the heels of legal psychedelic psychotherapy in the United States, which will give record numbers of Americans the chance to experience MDMA and psilocybin. Excitement has built around the empirical finding that depressed patients who received two doses of psilocybin reported durable attitude shifts toward a greater sense of connectedness to nature and away from authoritarian ideals, like xenophobia and ironfisted leadership.[2]

But are better attitudes enough to save us from the dual colossi of climate change and fascist strongman politics? We know that much of what comes out of a psychedelic experience is only as impactful as the integration that follows it. Our integration practices are what bridge the chasm between felt experiences and the changes they inspire. And, for the clinical protocols seeking FDA approval, this integration will be administered by an institution that has been seen by many as the natural home for psychedelics within American culture: Western psychotherapy. However, inadequate attention has been paid to the fact that psychotherapy is our culture's very particular take on how healing and change should happen—one that has co-evolved with our values and molded itself to a certain status quo. Our decision to wed psychedelics to it will thus have a profound impact on the ways in which we will allow them to change us. Psychotherapy is home to many biases worthy of critique, but perhaps the one most threatening to our hopes for a psychedelic transformation of society is its bias toward individual solutions in lieu of collective ones.

INDIVIDUALISTIC BIAS IN WESTERN PSYCHOTHERAPY

Psychotherapy is, at its most basic, a set of techniques for attending to suffering at an individual level. Even group or family modalities serve the well-being of the individuals present, not the larger social systems around them. This may seem like a trite, unproblematic thing to observe. But there was nothing fated about our choice to frame our dominant response to suffering in this way.

The roots of psychotherapy's individualistic bias can be traced back earlier than field itself, at least as early as the nineteenth century.[3] Americans were reeling from the alienation brought by the Industrial Revolution and the traumatic tears to our social fabric wrought by the Civil War. To fill the void left by our lost spirit of communal interdependence, self-help technologies arose, such as mesmerism, mind-cure, and the mental hygiene movement. What they had in common was that they taught Americans to reframe the adversity of the era as an internal problem with individual-level solutions. This mirrored the shift in societal power dynamics occurring at the time, as the newly ascendant class of advertisers and industrial managers was all too happy to help Americans learn that the one thing they could change was themselves.

These developments set the stage for the rise of American psychotherapy as a formal institution of individual healing for society's discontents. When psychoanalysis, the first major school of psychotherapeutic thought, arrived on our shores in the early twentieth century, its methods were quickly brought in line with our predilection for personal solutions. In Europe, psychoanalysis had been a disruptive cultural force whose revolutionary potential was a hot topic of debate among Marxists. On American soil, it quickly became a tamer set of self-improvement strategies that emphasized adapting oneself to a social context taken as given. Behaviorism, a homegrown contemporary of psychoanalysis, shared this tenet of changing an individual's response to conditions that were beyond their power to change. Humanistic and existentialist psychologies—the "third force" in psychology that arose in the 1950s—rejected the mechanistic language of their predecessors but further enshrined the individual as the domain for change.[4]

Multicultural competency training has become its own essential force within American psychotherapy, but the focus it has brought to cultural context has mainly served to help therapists attend to individual lives in a more contextualized way. Approaches that aspire toward collective transformation have also arisen in recent decades but have had little discernible impact on the way our culture does psychotherapy in its hospitals, clinics, schools, and private practice offices. The century of momentum behind the individualistic bias has proven to be too much to overcome.

So, for over a century, American psychotherapy served as our society's curator of individual-level solutions for many of our systemic problems. While it has helped many in this role, its more insidious fruits are very clear in the present day. Consider how many of our lethal epidemics—depression, suicide, addiction, gun violence—have systemic causes that get lost in our public conversations about increasing access to the panacea of mental health services. For these and many other issues, we have hamstrung our imagination for broader change by locating the root of our suffering within the individual.

If you were to approach 20 random Americans and say to them, "I am miserable, I feel alone, I am overworked, and I don't have what I need to better my situation," it would be striking if one of them raised the possibility of organizing collectively. The majority would likely ask, "Have you tried talking to a therapist?"

HOW THIS MAY LIMIT PSYCHEDELICS' DISRUPTIVE POTENTIAL

Psychedelics, coming to us by way of American psychotherapy, are set to inherit this individualistic bias. If they do, insights about our interconnectedness and relatedness to nature will be mined exclusively for the contributions they make to personal-level healing rather than considered as data that could drive the construction of a better world. In a word, they will be "psychologized."

For instance, imagine a person suffering from depression who undergoes psilocybin-assisted psychotherapy. In the session, he transcends his ego and finds his awareness expanded to include an entire ecosystem that is contaminated and on the brink of collapse. From this vantage point, he is able to experience its toxicity as something his human nervous system can understand: immense suffering. He shows up to his first integration session distraught about what he felt.

Within an individualist framework, his therapist may respond in a number of ways. She might see this experience as counter-therapeutic and try to talk him away from it. She may ask questions that draw his attention to the affect he's experiencing in hope that the deeper (read: personal) meaning will reveal itself. She may interpret it as an expression of inner conflicts, or early life material, as she would interpret a dream. Or maybe the moment will be taken as an opportunity to practice a distress tolerance technique. At best, the therapist may simply sit with him in the anguish, steadying herself in the faith that this is some mysterious part of *his* healing process.

What all of these responses have in common is that they take an experience about a large-scale injustice and make it about the patient. Such psychologizing is in line with what American psychotherapy has evolved to do. Psychedelics warrant a more expansive model of healing that leaves space for individual experiences to contribute to collective transformation. The direness of our current problems demands this as well.

LIBERATORY PRAXIS IN PSYCHEDELIC PSYCHOTHERAPY

Is there room for such an expanded notion of healing in American psychotherapy? In a word: almost! As mentioned above, the field's individualist bias has been increasingly recognized and critiqued in the past few decades by an array of dissident voices within its own walls, including feminist psychologists, Marxist psychologists, liberation psychologists, and critical psychologists. Many of their developments, if applied creatively, could point the way to a less individualistic approach to psychedelic psychotherapy.

As an example, among many, let's consider the idea of critical consciousness-raising. Radical educator Paulo Freire first developed this form of radical praxis, but we find its clearest elaboration for psychotherapy in the work of liberation psychologist Ignacio Martín-Baró.[5] Freire coined the term to refer to a three-step process of supporting individuals in becoming aware of the structural conditions they face, helping them build the self-confidence and motivation required to change these conditions, and co-determining how they might effectively do so. In Martín-Baró's approach, therapist and patient work together to recognize and unravel internalized forms of oppression in patients in order to turn them outwards to the world as agents of sociopolitical change who are "in charge of history." The desired outcome of this expanded take on healing is a patient who has not just healed their own trauma but has taken steps to uproot the conditions that set the stage for the traumatic event in the first place.

Now, imagine what the introduction of a critical consciousness-raising lens to psychedelic psychotherapy could look like. Imagine if the client who felt the pain of an ecosystem was invited to explore and confront the layers of culturally-enforced not-knowing that have hid his nature-relatedness from him and others? What if his therapist asked, not just about the benefits that this relatedness bestowed on him, but also the responsibilities? What if his therapist supported his post-session development in a way that held solidarity as equally important as individuation?

We would be acting on the opportunity to create the kind of healing that our world needs right now. And we would be doing justice to these awe-inspiring substances that we know can do so much more than "reset" a stuck brain.

Strikingly, this shift would require only a small, achievable step away from our current practices. It can be done now, and it can be done ethically. It need not come at the expense of the client's personal healing. If there were a conflict between an individual and a collective focus, the former would win out, but we are likely to find that they overlap more than they diverge.

There would be no added "pre-programming" of people's experiences with a demand for a collectivist focus.

The therapist need only to keep an ear out for any elements of the client's experience that may speak to collective concerns and to explore the client's openness to using it to benefit the collective good.

THE MUSHROOM AT THE END OF HISTORY

Perhaps this critique is too pessimistic, and psychedelics will burst free from whatever trojan horse they rode into our society to disrupt our dysfunctional systems. Maybe it is enough to simply "bring people to the medicine." I hope so.

But, if anything, I fear that the suggestion put forth here won't go far enough. Perhaps the old adage about the impossibility of dismantling the master's house with the master's tools applies here, and Western psychotherapy needs a deeper disruption from outside of its own walls. There are plenty of Indigenous notions of interconnectedness that could gainfully subvert our individualistic models.[6] There has been an inchoate call in the conference presentations of many psychedelic researchers to open ourselves up to Indigenous wisdom as our culture's use of psychedelics enters its turbulent adolescence. Maybe this is the lesson we most urgently need to learn at this juncture. The time is now, since, in the words of psychedelic psychotherapy-skeptic Terence McKenna, "we don't want this to end in a toxified garbage pit ruled by Nazis, but that is the way we may well be headed.... We need a pharmacological intervention on anti-social behavior or we are not going to get hold of our dilemma."[7]

Capitalism on Psychedelics: The Mainstreaming of an Underground

ERIK DAVIS, PHD

EVERYONE GETS WORKED UP ABOUT A SHOWDOWN, ESPECIALLY WHEN THE conflict involves colorful characters and positions you really care about. Like many attendees to the conference Cultural and Political Perspectives on Psychedelic Science, I was familiar with the *Statement on Open Science and Open Praxis with Psilocybin, MDMA, and Similar Substances*, which had been posted earlier this year on Chacruna and elsewhere.[1] The effort was spearheaded by Bob Jesse, a long-time member of the West Coast psychedelic intelligentsia, and a figure whose ethical, intellectual, and big heart bona fides are impeccable. Intervening in the rapidly developing field of psychedelic medicine, the Statement called for a continuing commitment to scientific integrity, data-sharing, and the spirit of service. It also reflected a growing distrust of the for-profit corporate behaviors that have recently been unleashed in the psychedelic space. Though Jesse was too careful to name names, the main sources of concern were two for-profit companies: Compass Pathways, a one-time nonprofit, now moving aggressively into the psilocybin therapy space in Europe, and Eleusis, a less visible but more cheekily-named outfit that has patented LSD for the treatment of Alzheimer's.

Scanning the list of names who had signed the Statement, I remarked again on something I noted in a somewhat cranky essay I wrote for *Erowid Extracts* in 2013: that the science of psychedelics cannot be disentangled from the wider and more multifaceted *culture* of psychedelics—very much including its underground culture(s).[2] My essay was aimed particularly at MAPS, who at that point had already established its current dominance over the space of "important" psychedelic conferences, all of which stressed the legitimating force of science in their titles and content. My main point was that if the "Multidisciplinary" in MAPS' name was going to count for anything beyond a groovy brand, the organization had to *actually* open the doors of the discourse it managed to disciplines other than science and clinical research. Though trained in the humanities myself, I was particularly concerned with the social sciences, which I hoped would provide some of the critical, contextualist, anthropologically-informed, and politically sophisticated correctives to the mainstream juggernaut of individualistic psychedelic pharmaceuticals already in motion.

MAPS has since come to open up more space for social science and plant medicines at its large and very successful conferences, and we are all better for it. Much of this has been the doing of Chacruna's own Bia Labate who, not coincidentally, was the guiding force behind the Cultural and Political Perspectives conference at CIIS. Here clinical and neurological discourses were pushed temporarily to the side, as conference presenters wrangled with a myriad of other questions, problems, and concerns that need to be addressed if psychedelic science is, again, going to be both genuinely multidisciplinary and fully engaged with the issues of social justice, representation, and political economy that are roiling our times.

Though the immediate purposes of the Statement for Open Science had little to do with social science proper, the document is also an important sign of the multiple communities and discourses that still make up psychedelic authority today. Though it was mostly signed by clinicians, scientists, and other researchers associated with institutions like Johns Hopkins, NYU, Heffter, MAPS, and Imperial College, the Statement also earned the support of underground elders, like Ann Shulgin, Kat Harrison, and Ram Dass (OK, Ram Dass got a PhD but only when he was Richard Alpert…). And though Bob Jesse played an instrumental role in the research at Johns Hopkins, he is not a credentialed scientist but a tech-head and organizer who also founded one of the Bay Area's longest-running spiritual dance collectives.

The reason I am highlighting the underground roots of some of these signatories has to do with the question of values. While the values expressed in the Statement align with Western science at its most ethical and idealized, the reality is that most "science" today is capitalism and technology in action: intellectual property rights, biased corporate-sponsored "studies," backscratching business-university couplings, and extraordinary rivalry for status and funds.

Moreover, the public rhetoric of "science" also continues to play a role in repressing perspectives—including Indigenous and "spiritual" points of view—that are not easily quantified or integrated into a reductionist framework. This is why, for example, MAPS conferences were once essentially restricted to presenters with degrees. In the world of institutional science, optics matter and the optics are restricted to the performance of licensed rationality.

And here's the thing about the psychedelic underground, for decades both science and R&D (research and development)—not to mention spirituality and healing—were practiced by outlaws. The institutional forces that dominate the knowledge and healing practices of modern society were kept at one remove (if not two or three), though the profit motive was not. Some of these underground individuals, but by no means all, made money producing and selling scheduled substances, and sometimes novel drug preparations. But profits alone do not capitalism make, and many of the writers, healers, freaks, and wizards who made up the underground, and who gathered at earlier psychedelic conferences, were centrally concerned with the values of the Statement: information sharing, mutual respect, service to the broad tribe of users, a disregard for the institutional rules of capitalism (but, again, not necessarily the market), all tied up together with a sort of Dionysian libertarian ethos that one might characterize as the integrity (and the pleasure) of thieves.

These somewhat idealized and nostalgic thoughts were stirred up recently when, in the weeks leading up to the CIIS conference, the DMT Nexus' David Nickles began using his human bullhorn powers to call out hypocrisy in some of the signatories of the Statement. Nickles particularly set his sights on Rick Doblin, who has been working in some still rather unclear ways with Compass, including strategy discussions, the coordination of training programs, and the possible joint sharing of research and study sites. (George Goldsmith, the head of Compass, has also described himself as a MAPS consultant.) While the Statement is non-binding, Doblin's cooperation looks like a double standard. It is true that other signatories to the Statement, including Johns Hopkins' inestimable Bill Richards, were also working with developing therapeutic protocols with the company, work that Doblin claimed Roland Griffiths supported.

Later, Jesse took the podium to correct this claim, pointing out that Griffiths has expressed reservations about cooperation with Compass.

By focusing on MAPS rather than Griffiths, Nickles kept the question of the values of the Statement (rather than narrower therapeutic outcomes) in play. The DMT Nexus, after all, is a paragon of underground science, a full instantiation of the outlaw research values describe above. MAPS, on the other hand, absolutely needs to visibly turn its back on this legacy to succeed. This has put the organization in the awkward—if nonetheless very successful—position of digging into the hearts (and pocketbooks) of underground psychedelic enthusiasts, while dragging everyone into the garish mainstream light of regulation, big money trials, insurance plans, professional agencies, and a normalizing discourse that requires the marginalization of earlier and more unruly psychedelic authorities, visions, researchers, and community values.

Nickles and Doblin appeared in a panel aptly titled *Capitalism's Systemic Issues: Will They Emerge in Psychedelic Medicine and Practices?* Bob Jesse began the panel with a cautious and sensitive presentation of his motivations in crafting the Statement. Without going into particulars, he expressed anxious concerns about the inherent dynamics of for-profit companies, who, in their mandated responsibility to shareholders, are required not only to bring the best products possible to market, but to "interfere" with potential rivals. Here Jesse was most likely thinking of the Usona Institute, a 501(c)(3) outfit, on whose board of directors he serves, which is designed to maximize the therapeutic potential of psilocybin at the lowest possible cost to patients. Imagining a world where both Usona's board and Compass's shareholders are grinning ear to ear is, needless to say, challenging.

Jesse's concerns were largely pragmatic, and keyed to an anxious but well-deserved understanding of how capitalism works on the ground. Rick Doblin followed up with more pragmatism, but from the other side. Given his perennially cheery view, which has never seen a conflict it can't ameliorate with a smile, it is unsurprising that, where Jesse sees the potential for destructive interference, Doblin sees a garden where a thousand flowers—including for-profit ones—may bloom. But Doblin has spent decades intelligently and idealistically navigating these regulatory and institutional questions, and MAPS' path to success is very solid. He discussed some of the ins and outs of intellectual property in regard to psychedelics, and outlined MAPS' anti-patent strategy and its attitudes towards data exclusivity, the sharing of protocols, and insurance machinations. He reiterated their plan to produce medicines through a Public Benefit Corporation that, using the relatively newish tool of benefit corporations, combines for-profit mechanisms with a commitment to social good.

Nickles' problem, however, is less with MAPS per se, but rather with who it considers its friends—deep divers into underground lore may recall the case of John Halpern.[3] In discussing Compass, Doblin did not address any of the larger, systemic problems with capitalism, pharmaceutical companies, or the current therapy regime. Nor did he address the distinct possibility that these mainstreaming forces may significantly corrode the global transformative potential for psychedelics at this tense juncture in the planet's history. This is because, as a pragmatist, he made his pact with "the System" decades ago. Doblin seems to genuinely believe that large-scale global transformation will occur by getting these substances into the nervous systems of people on an *individualist basis* under the current regime. What is important is the delivery of the medicine more than the motives of the deliverer. At one point, while discussing Compass, Doblin made an interesting verbal slip—though he wanted to say "they," he said "we" instead. Translation: for MAPS' global effort to succeed, even in its largely nonprofit terms, corporate players (and the shareholders behind them) are part of the team.

This is not a comfortable stance for those of us who believe that, as Nickles later put it, today's dominant culture is not a reality to be accommodated, but an existential threat to be resisted. Along these lines, Geoff Bathje followed up Doblin's presentation by repeating arguments against capitalism familiar to anyone who has been paying wide-screen attention to the political landscape in recent years. His strongest points concerned the way that psychedelic therapy, applied as "solutions" to individual psychological problems rather than broader social conditions, risks simply feeding into the self-improvement logic that increasingly underscores capitalist subjectivity, and that has already shown a remarkable capacity to absorb, defang, and redirect potentially transformative practices like yoga and mindfulness meditation. In this way, psychedelics may—and already are—contributing to the cancer that Bathje calls "elite perfectibility."

With firebrand fervor, Nickles followed up on these sentiments, but with the passion of an Occupy veteran. He drew attention to the inequities and iniquities of late-stage capitalism, asking those of us who consider ourselves responsible for the shepherding of psychedelics into the larger world to consider what we are signing up for in making a pact with the medical model and the "respectability politics" that accompanies it. He also took pointed digs at some of MAPS' more outrageous rightwing funders, like Rebecca Mercer, and argued that the fact that the notorious Peter Thiel has invested in Compass suggests, with apocalyptic clarity, that the mainstreaming of "psychedelic medicines" is already bound up with an authoritarian military-industrial-medical framework that does not have the world's best interests at heart.

Nickles is a radical, and Doblin is not. Indeed, it is rare to find such a diversity of political views in an academic panel. What are we to do with such divergence? The usual social media-abetted strategy is to choose a side and drill down on the enemy with hysterical intensity. But there is also a more ecological approach, one that recognizes that politics is a complex system, not a game of winner-take-all. As the human endeavors of psychedelic science and psychedelic culture continue to grow, bifurcate, and complexify, the field itself grows more multidimensional—and more genuinely multidisciplinary. Nickles drew necessary attention to many crucial issues with an understandable urgency, and this sort of critical work will inevitably pull and tug on other centers of power, forcing at least some of them to shift their practices (or at least their self-presentation). You don't have to agree in toto with his grim account of apocalyptic capitalism to recognize the positive and pragmatic effects that such rabblerousing may have on other organizations, thinkers, and companies as psychedelic science takes further steps toward the emerging cognitive regimen of psychedelic medicine. Efforts like the Statement, and Nickles' impassioned attacks, probably won't keep us out of the swamp, but they may help keep higher psychedelic values afloat.

This important panel was also a wake-up call. The "psychedelic community"—which no longer really exists as a singular entity, if it ever did—can no longer pretend that the process of mainstreaming is a purely positive, hope-for-humanity development that is separate from the larger crises of capitalism, militarism, authoritarianism, and the intensification of technological control over subjectivity. The values and motivations that once drove a strange, heretical pocket of the druggy world are now deeply woven into the larger, crisis-saturated world we all now claustrophobically share. The mainstreaming of psychedelics may well be a positive force in this chaotic and collapsing world, but it won't become so by simply smiling and shaking hands with the sharks that are already circling.

The very multidimensionality of the field may still be its saving grace. For all their social power, medical approaches will never entirely dominate the meaning of psychedelics, which also manifests in explicitly religious, recreational, productivity-enhancing, and intensely freethinking forms. As Jesse and Doblin both pointed out, the immense amount of written material that already exists about psychedelics—including by underground researchers—already precludes the range of possible patent moves by hostile or greedy players. And the continued existence of a robust black market, blockchained or otherwise, suggests that, while idealistic nonprofit approaches to aboveboard psychedelic therapy may have to fight it out in a marketplace more bitter and brutal than it should be, other avenues toward the world of the allies will remain open.

In the case of psilocybin, these avenues are widely distributed and painfully difficult to control. A few browser clicks will reveal cheap and easy routes of home production, supported by robust online communities along the lines of the DMT Nexus, to say nothing of the wide geographic distribution of shrooms growing in the wild, the eternal haunt of heretics and bored teenagers alike. This decentralized web of psilocybin production—so weirdly mirrored in the structure of mycelium itself—is all the ever-changing tribe of outlaws and freethinkers needs. Magic mushrooms are now in the light, but they will always grow best in shadow.

Note:

This paper provides a comment to the conference Cultural and Political Perspectives in Psychedelic Science, a symposium promoted by Chacruna and the East-West Psychology Program at the California Institute of Integral Studies (CIIS), San Francisco, August 18–19, 2018.

Profitdelic: A New Psychedelic Conference Trend

ASHLEIGH MURPHY-BEINER, MSC

ALMOST A DECADE AGO, BREAKING CONVENTION HOSTED THE UK'S FIRST-ever psychedelic conference at University of Kent (UK). Little did the organizers know, back then, that this would play a part in birthing the psychedelic renaissance in the UK. It was a small, unassuming event of 500 psychedelic aficionados who thought there was something valuable about these substances. Back then, anything to do with psychedelics was fringe and far from the mainstream.

Ten years after the first Breaking Convention, here we all are in a psychedelic movement that has changed beyond recognition. Suddenly, psychedelics are considered viable mental health treatments and they're popular. Michael Pollan's book *How to Change your Mind* has brought psychedelics into the homes of people who might otherwise have thought these substances were exclusively harmful. These are wonderful advances, but with this increasing popularity has arrived the potential for big business interest and the promise of psychedelic-fueled status, fame, and fortune—if you're willing to do what it takes to get it.

Perhaps it's no surprise that as psychedelics have become more mainstream, we've seen more conferences being created. The number of psychedelic conferences has been steadily growing over the last ten years, each one making a valuable contribution and becoming a welcome addition to the global community. However, in the last 6–12 months, we've seen a sudden and disconcerting rise of new startup psychedelic conferences. At first glance, these events almost look like the ones we know and trust. However, scratch just beneath the surface and you'll start to see the signs of profit-driven enterprises with little regard for the values or ethics this movement was built on. These might be better described as "profitdelic conferences." To understand how different these are in ethos and feel to what we've seen before, we can take a deeper look at what was driving the early psychedelic conferences.

Conferences, like Breaking Convention in the UK and Psychedelic Science and Horizons in the US, have become longstanding institutions among psychedelic communities and have played a vital role in advancing the academic, cultural, and political movements we see furthering psychedelic science and culture today. These conferences give people a chance to learn, a safe place to connect, and provide researchers from around the world with places to meet, get inspired, and forge collaborations. The kind of people who attended the first few Breaking Convention conferences were academic researchers, people treating their own mental health difficulties with LSD, ayahuasca, and psilocybin, or people who had discovered psychedelics as tools for exploring consciousness or spirituality. Most of the attendees had some personal desire to advance public debate about the safety and potential benefit of psychedelics. No one was there to make money, and this was a dramatically underfunded field of research. There was no immediate glamour, status, or fame associated with it.

The energy behind these pioneering conferences was one of advancing the fields of psychedelic research for public benefit and the movements towards legalization, or both. But even back at the first Breaking Convention, I was struck by an obvious tension within psychedelic communities that persists today. Most people want the freedom to use psychedelics legally but the thought of what legalized psychedelics might mean in reality has always been accompanied by fear. I remember a man in his 60's who genuinely and wholeheartedly believed in the potential of psychedelics standing up in the auditorium at the first Breaking Convention and saying, "Can you imagine what will happen if they're legalized in this culture? Do we want a Disneyland of psychedelics? I'd rather them stay illegal if that's the only option."

Perhaps he had a point; new players are now entering the field in rapid succession and, suddenly, it's no longer absurd to hear psychedelics and market forces spoken of in the same sentence. We are living in the era of psychedelic capitalism. Startup pharmaceutical companies are trying to patent psychedelic-assisted therapies, then the clinics funded by those companies are designed to distribute the medicines at a high cost to patients who can pay. These new pharmaceutical companies are also attempting to dominate the market by putting psychedelic researchers in key positions within their organizations. Additionally, legal startup psilocybin retreats are being set up by (it would seem) just about everyone with or without the experience to do so. The rise of profitdelic conferences is a part of this wider trend. These opportunistic conferences are being created by venture capital startups and people who make misleading claims about their experience or connections to the psychedelic community, borrowing credibility from the well-known psychedelic speakers they invite and using the speaker's legitimacy to recruit other speakers to fill up the conference and sell tickets. The content is rarely curated, and is more a mish-mash of anyone they could get on board who is known well enough and will sell enough tickets.

Speakers have started reporting issues with these conferences, as Dr. Bia Labate noted when she first highlighted and wrote on this topic.[1] A number of academic researchers recently withdrew their participation in an online conference. It was a cannabis industry conference for investors that included a psychedelic track for the first time this year to attract investors into the psychedelic healthcare market. Some researchers reportedly pulled out of speaking at the event over concerns the organizers had little knowledge of psilocybin for mental health issues, despite attempts to attract investors to invest in this, and that they had invited a speaker with no experience with using psilocybin as a treatment for people suffering with depression to be part of an expert panel discussing this topic.

Another profitdelic conference made claims about including Indigenous speakers, something which would lend credibility, but they never materialized in the final speaker lineup. The organizers offered attendees and speakers the opportunity to invest money in their startup psychedelic investment business. It was later discovered that, apparently, there was no psychedelic investment fund and the conference was an attempt to attract the funding needed to create it. Another issue that's been flagged with these kinds of conferences is that the organizers, lacking links to the psychedelic community, are unaware of the unethical and unsafe practitioners operating in this field. A number of these conferences have given a platform to psychedelic retreat owners or facilitators who are known within the community to be unsafe or unethical.

Despite these significant issues, these conferences keep popping up, securing lots of high-quality psychedelic speakers and selling lots of tickets.

The question is, how are profitdelic conferences getting away with producing these poor-quality conferences and getting people to present and attend? One factor is that most psychedelic researchers and practitioners aren't paying much attention to the events they're invited to. Up to this point, there has been little to worry about. All of the psychedelic conferences up to this point have had skin in the game and had been about advancing education and/or legalization. Psychedelic conferences have, more or less, been working towards similar ends, albeit in diverse ways. Until now, speakers have been able to have a quick glance at the website of the conference they're invited to when they get invited. Most will agree to speak at any conference that looks legitimate because they care about their work and want to share it.

Some other researchers and practitioners are starting to become aware of concerns about these conferences but still choose to speak there anyway. There are a few different reasons people give for this. For some early career researchers and practitioners, it's hard to say "no" to a speaking opportunity, even if they don't agree with the conference offering it, because they feel the need for the visibility and networking opportunities conferences offer, since they're trying to build careers in a field with scarce resources and jobs. For others, often more established in the field, there's the growing sense there's nothing they can do about psychedelic capitalism and its influence on the field, so they may as well go to these conferences and try to influence the conversation there. Some say they should join them in their capitalist pursuit and make money with them. There's also this idea bouncing around that unless every speaker refused to speak at these conferences, some people will always agree to speak, so not participating wouldn't make a difference anyway.

These are all valid and understandable things to think in an attempt to make sense of a confusing time in the psychedelic movements we're all involved in. So many changes so fast, and too many new people are entering the field to vet them all. But the concern is that, if we don't take a proper look at this and the long-term implications of what's happening now, this movement will slip through our fingers, turning into something hollow, completely losing sight of the values psychedelics first revealed. Now, it is more important than ever to ask ourselves why we got involved with this vast psychedelic movement, why we're still involved, and what we want it to bring into our world, our societies, and our relationships.

Every time we let something slip by that we don't think is right, we're moving further away from the values this movement was built on. Transparency, compassion for those suffering, challenging cultural norms, and an awareness of our interconnectedness have always been at the heart of it all. Waking up to what's happening now and making a conscious choice about the role we want to play in it is vital. We do all have an important part to play; each person's choices matter and will have an impact, however small.

It's important to note that for-profit approaches to psychedelics aren't inherently negative but, like any approach, they do need to make conscious their values and ethics to ensure profit doesn't supersede the intentions of the project and damage whatever good they set out to do in the process. A story titled *We Will Call It Pala* by The Auryn Project is a cautionary tale for the psychedelics of this.[2]

For the last few months, a few of us, including myself, Bia Labate, Anya Oleksiuk, Ros Watts, Mike Margolies, Lia Mix, Julie Holland, Pam Kryskow, Mareesa Stertz and an informal network of women called "Women and Psychedelics," have been trying to reach out to organizers and challenge motives, deficits in curation, and lack of diversity. This has been exhausting. We are realizing that, rather than trying to stop these conferences or give them credibility by collaborating with them or attending them, we might be better off focusing our efforts on supporting and building events we truly believe in—ones that embody values and ethics we can be proud of. At the same time, we need to keep raising awareness about this topic and this article is a modest attempt in this direction.

If you're thinking of attending or invited to speak at a conference, here are some ways you can assess the legitimacy of the event.

- Find out how long this conference has been around. A brand-new conference requires research to determine if you are supporting a legitimate event.
- Find out what the purpose of the conference is; don't just accept what

is written on the website. Find out who the conference is aimed at and what the personal and professional benefit is to the organizers for running the event.

- Google the organizers, find out who they are. How long have they been part of the psychedelic community? What are their backgrounds? What are their jobs?
- Ask if the conference is paying their speakers. Often, unscrupulous ones won't pay anyone even though ticket sales are high and it's a for-profit event. Good conferences should be transparent about remuneration.
- Look at the list of speakers to see if they make any attempt at diversity; many of the most well-known conferences still aren't meeting the best standards for this, but they have made significant progress in the past 10 years.[3] The new startups often lack diversity and create an echo chamber of limited wisdom.

Additional things speakers can do:

- Ask other speakers and conference organizers about the reputation of the conference organizers among people in the psychedelic community.
- Ask other speakers about their experiences of the conference you're being invited to and share your own experiences to help others make informed decisions.
- Support local grassroots psychedelic communities and organizations.

Considering the rise of profitdelic conferences, we need to create ethical, safe, and values-driven events and empower speakers in the field to do more due diligence and be more selective about where they speak.

SECTION FOUR

Queer

Queer Voices Speak to the New Psychedelia

JEANNA EICHENBAUM, LCSW

ON THE EVENING OF JUNE 1st, ABOUT 50 OR SO PARTICIPANTS AT THE QUEERing Psychedelics conference gathered in one corner of the Brava Theater in San Francisco to discuss what unique issues queer people might bring to psychedelic therapy for healing. As group facilitators, we felt honored and moved by the turnout. After all, it was 7 p.m. on a beautiful Bay Area day, and we'd collectively been in the space since about 9 a.m. that morning. Personally, I was surprised that so many decided this topic was important enough to stay after what was already such a long and full day. We learned much about the intersections of psychedelia and queerness throughout the history of the past 50 years, from the intense 1950s experiments and dialogues of Aldous Huxley and the probably queer Gerald Humphrey, through the early, dusty days of the Radical Faeries and Burning Man, and on to the glory days of The Coquettes and the early Black and trans club scenes of New York City. We listened to a critique of the limitations of mystical experience definitions when they are done solely through the lens of being White, straight, and male. Perhaps, most importantly, we were reminded of the original inhabitants and caretakers of the land we were sitting on, as a presentation of Ohlone consciousness and mindfulness opened up the conference.

After all this, I think we were expecting a small group, maybe five to 10 sturdy folks determined to add more voices to the mix. The group being so large and engaged was an acknowledgment of how important many queer folks feel this work is. It also exemplified how much needs to happen to make this LGBTQIA+ conscious and mindful to have actual healing potential for us.

The general sense of the gathering seemed to be one of sober reflection (yes, kind of surprising for a conference with this title!), the idea that our particular needs, fears, traumas, and resiliencies potentially bring so much to this new psychedelic renaissance, and a demand that our experiences be heard, reflected upon by the researchers and guides, and taken seriously.

One way of understanding the discussion is examining how we navigate various contradictions of experience to make use of these medicines in healing contexts. One of the early conundrums that arose was the idea that queer people are not used to being heard, and, perhaps more importantly, "gotten" by straight, establishment people and institutions. Given that, what do those people and institutions need to know to create "safe enough" psychedelic spaces for our communities? They need to know that much of our trauma comes from societal homo/trans/bi/pan phobia, and the resultant violence (physical, emotional, religious, and spiritual) that so many of us have had to endure and that attempts to address that trauma only within the confines of the office setting is not enough; a wider, holding, cultural and racial accountability is necessary.

Another contradiction that was discussed was the idea that the term "queer" itself is a loaded one for many people; a source of deep pride and galvanizing energy for some, and a painful reminder of bullying, beatings, and abuse for others. This also led to a number of discussions about how queer, or LGBTQIA+ people, are not one thing, nor one entity. We are actually many communities living in urban and rural areas with different races, classes, religions, educational and professional experiences, and expressions of intersectionality. Therefore, our needs are not a monolith but need to be assessed and honored individually, even while we are a movement for full rights and healing for all.

A third conundrum arose around the differences between the ideas of safety and comfort. We, as queer people, need psychedelic spaces that are "safe," meaning: free from abuse—sexual, spiritual, and emotional—with sensitive and informed practitioners who both mirror our experiences as Black, Brown, Native, Asian, older and younger, and urban and rural queer people. But deep healing often calls for and necessitates encounters with aspects of ourselves that are particularly uncomfortable, including working through queer-bashing and family-of-origin rejection.

Another potentially uncomfortable but rich topic that could arise in psychedelic sessions is that queer people might be drawn to explore the ways families of origin didn't "get" them, or weren't the "right fit" for them, because of their family's heterosexuality. That this lack of fit is often a deeply traumatic experience that LGBTQ+ people endure. Of note is the idea that this "lack of fit" might, in and of itself, be regarded as a kind of initiation—an initiation into other spaces, other ideas of family, connection, and community beyond the nuclear family that is perhaps not available to our straight brethren. So, the question of how to create the scaffolding of safety, honor, and respect that actually allows us to confront and heal trauma and allows the fullest possible exploration of the multiplicities of our selves was an important talking point.

Some further questions and ideas that arose included the following:

1. What does a queer spirituality that includes the multiplicity of our various identities look like? One that honors and includes sex and the body in so many of their healing and pleasurable manifestations; one that honors those on religious paths, and those who find those paths too historically painful, or actually irrelevant, to be necessary; one that acknowledges the loss of so many of our potentially wise elders and teachers during the plague years of the 80s and 90s and the need for that wisdom now.
2. What forms of therapy (cognitive, psychodynamic, narrative, etc.) might be most useful to allow us to define and tell our own stories in the ways they need to be told?
3. What types of guide configurations outside of the limiting male/female dyads might be most helpful for us, and what types of healing settings could be best? Are group sessions a better way to increase trust and build healthy communities than one-to-one or two-to-one client and therapist sessions?
4. Honoring the centrality of the land and the ancestors of this land that we are privileged to be doing this work on; looking for meaningful ways to bring awareness of the history of the land and the First Peoples who were, and still very much are, here as we engage in our healing.
5. Inviting participants in these healings to bring the objects and artifacts that are meaningful to them for altar placement as both important in its own right, and one of the ways we consciously create the old/new queer healing rituals that will flow into the future.

There are two ideas I want to offer in closing. First is the awareness and acknowledgment voiced by several people that, as queers, we are particularly adept at traversing liminal spaces—the "in-between," not yet arrived, or always-on-the-way. This is a particular gift that might be especially resonant in work with psychedelics, which, in themselves, point to and inhabit those spaces. And, lastly, as several have noted, the plants and medicines themselves have their own intelligence, no doubt much vaster than ours, and have arrived with us here at this particular moment in time (as they have been here for thousands of years in dance with other cultures and environments), perhaps for reasons we cannot know or understand, but which we must simply accept and learn from.

Psychedelics Are Queer, Just Saying

BETT WILLIAMS

I ATTENDED THE CULTURAL AND POLITICAL PERSPECTIVES ON PSYCHEdelic Science Conference wearing a dirty cowboy hat, in keeping with my self-conscious awareness that I was an imposter, being that I am a non-academic writer, an outlier tripper, and former mushroom grower (for personal use) from an old coal town in New Mexico—a true psychedelic vortex on the map. It's a place full of people so eccentric that being a lesbian is about as controversial as saying you like cheese. The panel I was moderating was called Psychedelics and Sexual and Gender Minorities.

I arrived with the question, "Are psychedelics queer?"

I think, yes. I am using the word "queer" rather than LGBTQ+ etc., because, like psychedelics, queerness exists in a realm larger than identity politics. Queer is a realm of being that is personal, political, spiritual even, and like psychedelics, the form it takes is dependent on cultural context.

Queer has always been whatever we decide it is.

I wondered if queerness even mattered in the realm of psychedelic science, an arena currently fraught with issues of controversial corporate funding—MAPS' acceptance of a donation of one million dollars from the Mercer Foundation being one example, and Peter Thiel's creation of Compass Pathways, the company behind the promotion of pharmaceutical psilocybin, another. We are dealing with structural institutions that, whether they mean to or not, foster hetero-White male dominance from the get-go. Yes, Peter Thiel is an out gay man, but, whatever. This is exactly my conundrum, and one I brought with me to the conference. Is queerness really psychedelic? If so, does that mean it's an innately a revolutionary force?

I connected with the panelists at lunch and my ambivalence about the question faded quickly. I was amongst familiars who clearly had done serious time in psychedelic space, both on the ground politically, in their careers, and in the zones. I was taken in by the vulnerability of Jae Sevelius, who was quite nervous about her presentation. Jae works with the Center for AIDS Prevention Studies (CAPS) and leads several research projects at the Center of Excellence for Transgender Health. She was no novice to speaking in public, but she wasn't used to having to communicate a strong opinion, and yes, this time, she had one. She described her reaction to MAPS accepting the Mercer Family Foundation donation as one that "floored and appalled" her.[1] It was a gut feeling that I could tell was still with her as I watched her pull herself together before she took her place at the podium in an attempt to hold back a flood of emotion.

"Should psychedelic science be concerned with social justice issues?" she said, with an eloquent and arresting composure that held the room.

She went on to say right action is not about enforcing "diversity" in psychedelic spaces. It's about bringing the values of these marginalized communities to the conversation, integrating those values, and, if possible, centering marginalized groups and giving them actual power, like real jobs, publishing their views, and monetary support.

There's diplomacy that arises in times of crisis and great change simply because it has to. When political strategy comes with an open heart mixed with intellect, a track record of activism, and genuine curiosity, it's an alchemy that can turn golden, kind, and joyous. Jae brought this energy to a room full of people equally capable of guiding this new wild critter show we call the new psychedelic revolution to a better fruition than what's been writ for us thus far.

All this while Rick Doblin himself was sitting in a foldout chair to her left.

Gregory Wells, PhD, is a psychologist who specializes in psychedelic integration. Via his genuinely fierce slide show on the old San Francisco LGBTQ+ scene, he laid out exactly what queers have brought to psychedelia. He gave a brief history of The Cockettes, an avant-garde theater troupe influenced by the films of Jack Smith, and yes, a ton of LSD. The Cockettes influenced club culture, the Sisters of Perpetual Indulgence, the Radical Fairies and as a result, nearly every fundraiser ever held at a gay bar or pride parade. It was never just about a good party. The queer culture Wells speaks about has always been serious about people staying alive—and giving us all a reason to want to live. The resurrected bar, The Stud, a psychedelic wormhole in gentrified San Francisco, has a direct lineage to this psychedelic tradition.

Clancy Cavnar, a visionary artist, ceremonialist, therapist and writer, stepped up and put a necessary dark spin on how LSD was used in the early days by the medical and psychological establishment as a form of conversion therapy. At lunch, we all had spoken with levity about how Ram Dass (Richard Alpert) used LSD in hopes it would turn him straight. I had no idea, however, the level to which this was a "thing."

"Stanislav Grof used LSD to treat homosexual clients," she said. He came away with the usual cliché theories prevalent at the time. Gay men were afraid of the Vagina Dentada. Lesbians just wanted to return to the mother. Grof noted that homosexuals, "saw their sexuality in archetypal or transcultural ways, such as witnessing fertility rites, initiation ceremonies, and temple prostitution."

I try to picture these poor gays in Grof's therapy office, trying their best to give him something worthy to chew on and coming up with elaborate visions resembling scenes in *The Wicker Man* or anything by Alejandro Jodorowsky. Set and setting, indeed. My most lesbian mushroom trip I ever had was when my dog Spanky told me she liked the chicken dog food more than the lamb. "You like lamb," she said. "I like the chicken."

I suggested to the panelists that psychedelics are queer because you get to "be everything." Even psychedelic people who identify as straight I'm sure can relate to this. My question is how actual queers affect psychedelic culture as a whole. The panelists had no ambivalence around this question.

Queers know how to make community in times of crisis. We know how to mobilize outreach in larger circles. We set up fundraisers, community centers, nightclubs, recovery rooms, galleries, and non-profits with all of the skill, good graphic design, pleasant attitude, and finesse of Dolly Parton on Modafinil. Queers used to be illegal, just like the drugs we like. It doesn't feel very good. We don't like it. Gonna change it. Queers are as good at getting addicted as we are at getting free. We know how to party, but we need each and every one of us here. The dating pool is small enough as it is, so we work hard at helping people recover from addiction and other traumas so we can keep seeing them around. Even people we hate, we like seeing around. Queers did PR for safe drug and alcohol use way before Burning Man built their chill out/med tents, way before Noah Levine capitalized on substance recovery for outsider types. Lastly, queers seem to know that a broken heart can be a crisis as serious as a car crash.

To watch the psychedelic mainstream move forward with the medicalization of psilocybin and MDMA, it's hard not to remember the AIDS era, when drugs were being withheld for reasons having to do with politics, bigotry, and corporate capital. They were finally given to us, but at a price. I haven't forgotten. We were supposed to say thank you when they finally decided gay, trans, and BIPOC lives were worth saving. Unlike AIDs drugs, psychedelics are relatively easy to acquire and administer. Herein lies my dilemma with psychedelic science and the corporate pharmaceutical machine. How deep do we go to lead our communities towards mainstream legitimacy? Should we jump ship before the dominant system gets even more ugly?

I learned from the panelists that I don't need to view it so much as a black and white issue. That it's possible to live in many worlds—another thing us queers do well. We can live a rogue outlaw life while at the same time working to change the mainstream psychedelic culture with the intent that maybe we can make it less gross, maybe even fabulous.

Clancy's talk about LSD's use in conversion therapy could have been just a fascinating journey through a dark history, if it wasn't for this one fact she presented:

"In 2016, for the first time, the Republican Party endorsed conversion therapy in their platform under "right of parents to determine the proper medical treatment and therapy for their minor children."

Conversion therapy, or "reparative" therapy, is illegal in only 16 states, and has been promoted in the past by current Vice President Pence, who said, "Resources should be directed toward those institutions which provide assistance to those seeking to change their sexual behavior,"[2] though he currently denies his support.[3]

I think of a New Age workshop I attended at age 15 in Santa Barbara. An entire room full of people emanating "light," tried very hard to convince me I was bisexual, not a lesbian. The whole ordeal ended with me in a tearful panic attack that thankfully turned to rage. I screamed, "Do you have any idea what it is like to come out as a lesbian at age fifteen?" It was 1985. Had that homophobic attack happened in psychedelic space, I shudder to think of the possible psychic wound that might have calcified within me. We are queer and we find our way. It's easy for me to forget how bad it can get, how tragic it can be, and was, and is, for some.

As a therapist, Jeanna Eichenbaum works deep in trenches of the LGBTQQI community, offering trauma therapy as well as psychedelic integration support, which she has been trained in extensively.

She explained the use of the word "queer" in English language: peculiar, eccentric, strange, unusual. She stole my heart when she said, "It's an odd word which takes us, in ostensibly one syllable, through at least three distinct movements of the lower facial muscles. That is queer!"

She said one of the qualities of religious experience as described by William James in *The Varieties of Religious Experience* is, "A feeling that one has somehow encountered "the true self" (a sense that mystical experiences reveal the nature of our true, cosmic self: one that is beyond life and death, beyond difference and duality, and beyond ego and selfishness."

Before Jeanna came out as the woman she is, she spent many years struggling with her identity, in a state of horrible self-hatred. She had no friends and was living a life in secrecy and hiding. She met a man who claimed to be an experienced psychonaut. He was convinced that if Jeanna took a large dose of LSD or psilocybin, she would work through her issues and figure out she was, indeed, a man. Longing for a "normal" life, Jeanna agreed to trip with him. This prayer for "normal" began with 400 micrograms of LSD.

It didn't go quite as her guide had planned. Basically, what happened was this: God harangued Jeanna for quite a while, for, like, an eternity, asking her over and over, "WHAT DO YOU WANT???" In the face of the voice's insistence, she began blurting out everything she had ever been ashamed of until finally she was empty, done, and exhausted. Lots of other crazy things occurred that only Jeanna has the talent to tell. It was epic. Deserts and ocean waves were involved.

Finally, God asked her, "Is there anything else?"

"I paused, swallowed," Jeanna said, "I said, yes, there's one more thing. I want to be a woman. There was a long pause. I could hear the desert wind rustling through the cacti. I could hear nothing, the long, empty moan of nothing. Then, suddenly, in a singsong lilt, the voice playfully asked, with a note of the trickster, 'What's wrong with that?'"

Can Psychedelics "Cure" Gay People?

CLANCY CAVNAR, PSYD

ALTHOUGH IT IS COMMON FOR PEOPLE CURRENTLY TO EXPRESS THE OPINion that psychedelics will open people's minds, making them less judgmental and more able to share love and be inclusive, one early use of psychedelics by psychologists was in attempts to treat homosexuals to change their sexual orientation. "Conversion therapy" is a treatment that attempts to change the sexual orientation of homosexuals. This practice became more controversial after the removal of homosexuality from the DSM in 1973.[1] If homosexuality is no longer an illness, what justification could there be for treating it?

In 2016, for the first time, the Republican Party endorsed conversion therapy in their platform under "right of parents to determine the proper medical treatment and therapy for their minor children." Conversion therapy, or "reparative therapy," is illegal in only 16 states, and has been promoted in the past by former Vice President Mike Pence, who said, "Resources should be directed toward those institutions which provide assistance to those seeking to change their sexual behavior," [2] though he currently denies his support. [3]

Freud believed all humans were born bisexual, and that their later preferences were the results of life experiences and conditioning from parents. "Homosexuality is assuredly no advantage, but it is nothing to be ashamed of, no vice, no degradation, it cannot be classified as an illness... It is a great injustice to persecute homosexuality as a crime, and cruelty too."[4]

Kinsey also normalized homosexuality by creating the "Kinsey Scale" and expressing sexual orientation on a continuum. However, in the 50s, with the advent of behaviorism, the idea that homosexuality was a learned behavior was introduced, and so treatment to cure it with behavioral interventions, including aversion therapy, the pairing of a painful event with the behavior that is desired to be extinguished.

The opinion of the American Psychological Association (APA) is that there are no safe or effective ways of changing someone's sexual orientation, and therapies that claim to do so can reinforce negative views of homosexuality and be harmful to the client.[5] The ethical foundation for treating individuals with socially undesirable traits with powerful psychedelic compounds to rid them of these traits is highly suspect. This article is a reminder that all medicines can be poison in the wrong hands.

SEXUAL MINORITIES AND TREATMENT WITH PSYCHEDELICS

A study by Alpert was one of the earliest reports in the literature on sexual minority experience with psychedelics.[6] It is one example of conversion therapy found in the literature (see also, Masters & Houston, 2000; Martin, 1962; Stafford & Golightly, 1967). Alpert administered 200 micrograms of LSD-25 to a male self-identified bisexual volunteer who was dissatisfied with his attraction to men. During his fifteen-hour trip, the subject was shown pictures of women and encouraged to develop feelings toward them. In subsequent LSD sessions, a woman the subject knew was present and he had sexual intercourse with her. One year after the treatment, Alpert reported that the man was living with a woman, but he had two subsequent homosexual encounters, which the subject described as tests of himself to see if the changes he had experienced because of the treatment were "real." Alpert explained that the use of LSD allowed the subject to take a broader view of the archetype of "woman" and find connections to primal desires within the archetype, which he could then generalize to all women.

Stanislav Grof treated homosexual clients with LSD.[7] He concluded that gay men's dislike of sex with women was related to images of "vagina dentate" and castration fantasies that were envisioned during LSD sessions. He related lesbianism to the desire to be close to the mother. Grof admits that he has treated mostly homosexuals who were dissatisfied with their orientation, and that a healthy adjustment to same-sex orientation is possible and may not represent intra-psychic struggle. Grof also noted that subjects in LSD treatments often saw their sexuality in archetypal or transcultural ways, such as witnessing fertility rites, initiation ceremonies, and temple prostitution.

In *The Varieties of Psychedelic Experience*, Masters and Houston reported giving the psychotropic cactus peyote to a group of gay volunteers.[8] Working from the assumption that homosexuality is an undesirable orientation, Masters and Houston attempted to treat their gay clients with repeated doses of peyote. They reported that 12 out of 14 male homosexual volunteers in a psychedelic experiment had distorted body images that the researchers contended to be causal to homosexuality, although they admitted they could not prove this. They found that by taking psychedelics, the body image distortion was corrected, and they observed a trend toward "heterosexualization." They also speak stereotypically of the passivity of the homosexual man being transformed by psychedelic therapy, and attribute a deepening of the voice, greater vigor, improved posture, and greater masculinity to treatment with peyote. They found that the participants displayed a greater desire to appreciate their appearance after the peyote experience. In one case reviewed by Masters and Houston, the researchers were discouraged by their subjects' "considerable investment in his homosexuality" and felt unable to capitalize on the "gains" made in therapy.[9] They proceeded to speculate on the progress they might have made had the subject been more motivated to become heterosexual.

A study by Martin looked at the effects of LSD on 12 gay men.[10] Martin recommended LSD as a treatment for homosexuality. Administering many low doses in a treatment known as psycholytic ("mind-separating") therapy and encouraging intense mother-transference, Martin claimed that seven out of 12 achieved heterosexual orientation with only one "slight relapse" in a 3- to 6-year follow-up.[11]

Stafford and Golightly reported on LSD therapy for homosexuals during the 1960s. They found that homosexual issues were often resolved using psychedelic therapy and that homosexuals would either be at peace with their orientation after LSD therapy or decide that they were heterosexual.[12] Stafford and Golightly viewed homosexuality to be a result of early childhood trauma and "morbid dependency" on parents, both of which could be treated with regressive "shock therapy" with LSD. Stafford and Golightly recommended that LSD be used to treat transvestism, fetishism, and sadomasochism in the same way that it could be used to treat homosexuality. Masters and Houston also advocated LSD, stating, "Treatment of sexual disorders—frigidity, impotence, homosexuality and fetishism—and some other neuroses has many times been described as both drastically shortened and made more effective when LSD was used as an adjunct to psychotherapy."[13] This view reflects the thinking current in the late 60s, in which homosexuality was viewed as a mental disease, related to paraphilias.[14]

PERSONAL ACCOUNTS

Rare personal reports of individual therapeutic entheogenic experiences of sexual minorities can be found in the past literature prior to the current psychedelic renaissance. First-person stories about self-administered entheogenic experiences by gays and other sexual minorities help expand the story. Some examples from before the latest psychedelic renaissance include are port made by a "post-op" transgender woman describing her most recent experience using LSD.[15] She and a "pre-op" transgender woman agreed to take LSD and, at the peak of their experiences, they agreed they would look at themselves naked, side-by-side in a full-length mirror, "We would look to see whether we were monsters or whether we were God's beautiful creatures. And through the wide open doors of perception, we saw the truth: We were beautiful."[16]

Berkowitz, a lesbian, wrote about encountering her grandmothers in a vision during an ayahuasca experience on her 30th birthday. She concluded her report by saying that she felt she had "reclaimed (her) life."[17]

Merkur described a man who took LSD and was able to integrate and accept the fact that he had had homosexual experiences in the past, experiences he had previously been unable to reconcile with his self-image. He came to the conclusion that it was not a "big deal"[18] and during this experience, he was able to perceive himself in a non-judgmental way that proved healing for him.

Annie Sprinkle, a bisexual sex worker, educator, and performer, wrote about her experiences with drugs and entheogens. She had not tried ayahuasca, but had taken "pharmahuasca," a combination of chemical and natural sources of DMT and MAOI. She related that she felt that experience was preparing her for her death.[19] Her experiences lead her to the conclusion that entheogens can have a role in sex therapy because they can help individuals gain a fresh perspective on their identity. She posits that sexuality and the use of entheogens are both about consciousness and self-discovery.

These and other examples throughout time, including my dissertation work on gay people and ayahuasca, show that gay people can benefit from psychedelic use to heal from cultural homophobia. Psychedelics are rightfully used to "manifest mind," not to warp it, as these efforts at conversion therapy and other reckless uses, such as in MK-Ultra experiments, reveal. It is important and timely to raise awareness of how agendas may color the conceptions and uses of psychedelic medicines; a relevant topic as the market begins to form around psychedelics once again.

10 Calls to Action: Toward an LGBTQ-Affirmative Psychedelic Therapy

ALEXANDER BELSER, PHD

COMBATING HETERONORMATIVE PARADIGMS IN PSYCHEDELIC SCIENCE

To my friends in the psychedelic community: we have rainbow skeletons in our closet. Many people do not realize psychedelic medicines were historically used as "conversion therapies" to treat lesbian, gay, bisexual, transgender, and queer people to become cisgender heterosexuals.[1] Psychedelics were employed as tools of oppression targeting LGBTQ+ people in anti-therapeutic ways. We have yet to reckon with the historical harms inflicted on sexual minority people, as many of those harms continue to be perpetrated today. It is the aim of this short article to educate about our past failings, describe our current shortcomings, and offer constructive actions for the future.

The history of psychedelic homophobia is replete with harrowing stories of LGBTQ+ people targeted for treatment of "sexual perversion." A recent article in *The Lancet Psychiatry* reports that women in the United Kingdom were treated with LSD to "overcome their sexuality."[3] We also know from published case reports that large doses of LSD were used in conversion therapies to treat patients including: "one transsexualist, one transvestist, five male homosexuals, and one female homosexual."[3] The conversion clinicians boasted success in turning gay men straight. For example, after being administered 200 micrograms of LSD, one homosexual male patient refrained from homosexual activity for eight months following treatment. The researchers applauded his progress, reporting that he "felt only once the impulse to indulge in frottage and that only fleetingly. He is much happier, relaxed and contented and claims to be better than he has ever been since his perversion began."[4]

The misuse of psychedelic medicines to treat "sexual perversion" occurred on both sides of the Atlantic. A recent study discovered accounts of 12 men with same-sex attraction who were treated with LSD-25 at Hollywood Hospital in British Columbia, then a preeminent international psychedelic center that treated Cary Grant and others. Treatment with LSD often failed to produce a "satisfactory heterosexual adjustment."[5] Masters and Houston, in contrast, claimed that treating more than a dozen gay male volunteers with peyote led to "heterosexualization."[6]

These are not marginal cases. Throughout the 1960s, the psychedelic movement was saturated with profoundly homophobic thinking, and respected luminaries publicly touted LSD's ability to make you straight. Timothy Leary proclaimed, "The fact is that LSD is a specific cure for homosexuality."[7] Ram Dass, then known as Richard Alpert, also used psychedelic medicines to encourage gay men to become heterosexual. Alpert published accounts in which he claimed treatment with LSD helped his "homosexual" patient stay in a long-term relationship with a woman and that "they have been having intercourse every night."[8] Alpert himself was closeted at the time and later spoke publicly about being attracted to men.

Of course, these studies must be understood in the historical context in which homosexuality was considered a disordered pathology. It was not until 1973 that homosexuality was removed from the *Diagnostic and Statistical Manual of Mental Disorders (DSM).*

Today, conversion therapy is considered to be unethical, anti-therapeutic, and profoundly harmful to those who undergo it. The American Psychological Association Task Force reported risks involved "confusion, depression, guilt, helplessness, hopelessness, shame, social withdrawal, suicidality, substance abuse, stress, disappointment, self-blame" and other adverse effects.[9] Such practices have been widely outlawed or condemned by professional organizations, including the American Medical Association, the American Psychiatric Association, the American Psychological Association, the National Association of Social Workers, and the American College of Physicians. I have frequently treated survivors of conversion therapies in my practice as a psychologist, and it is painful to witness firsthand this type of psychological harm.

These are the rainbow skeletons in our closet. This is our shameful past. All of this history has been buried, and it is time to unearth it. Yet, there has been almost no meaningful discourse in the psychedelic research community about it. Psychedelic researchers and therapists in academic settings continue to proceed without examining their straight and cisgender privileges.

For example, we continue to require "male/female" therapy dyads in many clinical trials, which revert to traditional gender role stereotypes, essentialize gender, and instantiate a gender binary that disenfranchises trans* people. The male/female dyad presumes that the reproduction of a mother and a father in the form of a heteronormative family is a valuable and valid tradition, as though it offered a comforting sense of safety for all. Additionally, despite recent progress by MAPS, many psychedelic clinical trials still do not assess for sexual orientation, gender identity, preferred name, or pronouns on intake. None of the quantitative psychedelic research in the last 15 years has published on the differential effects of psychedelic treatments as they apply to sexual and gender minority people—cisgender straight norms are assumed throughout.

This must be remedied. We can do better. The only way to dismantle non-affirming practices against LGBTQ people in psychedelic research is for all of us to take proactive steps on every level. We must actively work to not propagate heteronormative paradigms.

Here are 10 steps we can take as a community to systemically root out homophobia and transphobia in our policies and practices.

1. Confront structural heterosexism and transphobia at every level.
Address structural factors in psychedelic research: the predominance of White straight cisgender male researchers, funding structures, research priorities, and lack of LGBTQ+ patient-centric approaches.

2. Retire the male/female dyad and replace with gender neutral dyads.
The male/female therapist dyad is gender essentializing. Move away from gender stereotyped role assumptions. Move toward gender-neutral and client-specific approaches, so that therapist pairings are made on the basis of individual clinical profiles, trauma histories, and gender preferences.

3. Acknowledge that sexual minority stress causes disproportionate harm to LGBTQ+ people.
Sexual minority individuals have twice or greater the average rates of depressive, anxiety, and panic disorders, as well as problematic alcohol use as compared to heterosexual peers. This is not due to sexual identity, but rather to the extra stress that LGBTQ+ people face on a regular basis throughout their lives. This must be acknowledged in our therapeutic work.

4. Research the question of differential responses between LGBTQ+ individuals and a heteronormative population.
Assess sexual orientation and gender identity in all psychedelic studies. Run subgroup analyses to evaluate treatment response for sexual minority people. Run meta-analyses across studies to aggregate data for LGBTQ+ populations.

5. Create new affirmative therapies by, with, and for LGBTQ+ people.
Develop and test LGBTQ-affirming psychedelic-assisted psychotherapies (QA-PAP). The new approach would be guided by evidence-based principles of affirmative psychotherapy.[10] A model pathway is provided by Dr. John Pachankis.[11] Development of an LGBTQ-affirmative psychedelic treatment would consist of three stages:

(a) qualitative research, interviewing LGBTQ+ patients and expert mental health providers

(b) treatment development and pilot trial

(c) randomized-controlled trial of QA-PAP vs. a generic psychedelic-assisted psychotherapy. Identify unique LGBTQ+ stressors and mechanisms.

6. Do psychedelics lead to any change in sexual orientation? Do they make straight people less homophobic? Does psychedelic experience increase self-acceptance?
Despite the history of conversion therapy, many people have shared that psychedelics were helpful in exploring their gender and sexuality, and the psychedelic movement is intertwined with the sexual freedom and gay liberation movements. We should assess sexual and gender identity, attraction, sexual behavior, and internalized homophobia in all participants, pre- and post-treatment in all psychedelic studies. How do people change in regard

to acceptance of repressed identities? Do reported sexual or gender identities change? Do gay individuals become less transphobic?

7. **Pursue sex and sexuality research.**
Incorporate the body, sensuality, sex, and sexuality in our research and praxis. Our current paradigms seem very sex-phobic.

8. **Ally and get intersectional.**
Address how our multiple identities (e.g., sexual identity, gender identity, race, ethnicity, class, Indigenous status, immigration status, disability) affect each other (intersectionality). Ally with other target groups to dismantle pre- and post-colonialist patriarchal, homophobic, and transphobic structures in ayahuasca circles, underground work, and clinical research.

9. **Access queer wisdom.**
The psychedelic community can learn from queer people because our wisdom is intersectional; we come from every family, every walk of life, every religion, and every path. Our lineage is pluralism—we are by nature a convergence.

10. **Queer the "Mystical Experience."**
Queer spirituality, like psychedelic spirituality, can be embodied, relational, political, and visionary.[12] Reform psychedelic spirituality research from its narrow lineage.

The psychedelic practices of the past may have been homophobic and transphobic, but our future need not be. It's time for a reckoning with the heteronormative history of psychedelic research. Let's join together to address and dismantle the subtler forms of cisgender heterosexual male privilege in our policies and practices today. I believe we are up to the task.

Why Oneness Is Not Incompatible with Identity Politics

KATHERINE A. COSTELLO, PHD AND
MARCA CASSITY, BSN, LMFT

"WE ARE ALL ONE" IS THE DOMINANT NARRATIVE ABOUT PSYCHEDELIC TRUTH AND HEALING

The belief that the ultimate psychedelic experience and truth that psychedelics have to teach is oneness isn't just held by those psychonauts who oppose equity initiatives. It is the dominant narrative in both psychedelic research studies and the broader psychedelic community.

Many clinical research trials use the Mystical Experience Questionnaire (MEQ) as a way to try to understand and predict psychedelics' effectiveness in relieving depression, anxiety, end-of-life dread, and other symptoms. The working hypothesis is that if people have a mystical experience as defined by the MEQ, then they will have a reduction in symptoms. While the MEQ also measures positive mood, ineffability, and transcendence of time and space, under the mysticism category, 6 of the 15 measures are related to oneness, with the others focused on the sacred and the noetic. The oneness related measures include "Experience of the fusion of your personal self into a larger whole," "Experience of oneness or unity with objects and/or persons perceived in your surroundings," and "Experience of the insight that 'all is One.'" Oneness is described as an experience of sameness, a merging with others into a greater singular whole. In the psychedelic community, "We are all one" most often stands in for "We are all the same," a point we return to below. The issue isn't that the MEQ measures these experiences but that it defines the mystical experience almost exclusively in their terms.

As Dr. Alexander Besler explains in his lecture, "Queering Psychedelics: A Queer Critique of the Psychedelic 'Mystical Experience'," it does not measure other experiences that may also have something to do with healing outcomes, such as encounters with spirit guides, powerful somatic-based energetic experiences, reparative relational experiences, connections with nature, and strong sensory or even sensual experiences. According to the logic of the MEQ, if a patient has not had an experience of oneness, then they have not had a mystical experience, implying they have missed the healing potential of psychedelics.

This privileging of oneness is also pervasive in the wider psychedelic community. It finds its origins in Aldous Huxley's seminal 1954 *The Doors of Perception*, which influenced Timothy Leary, who, in turn, helped shape an entire generation's discourse on psychedelics, the effects of which are still tangible today, no matter how hard the community has tried to distance itself from this controversial figure.

THE EXPERIENCE OF ONENESS MAY BE CORRELATED WITH PRIVILEGE

This pervasive narrative that psychedelics reveal "we are all one" and that the healing gained from them is tied particularly to that experience is potentially harmful to all, but especially to marginalized communities, and even more so to the LGBTQIA2S+ community. Any discourse that privileges a singular narrative runs the risk of being oppressive since it inherently delegitimizes other narratives. Thus, the trope of oneness as the ultimate psychedelic experience leaves no room for other experiences of psychedelics that may be equally meaningful and healing. Equity in psychedelics requires taking seriously the multiplicity of experiences that patients and users have. If it does not, psychedelic-assisted psychotherapy risks becoming re-traumatizing instead of therapeutic.

Indeed, part of the trauma of LGBTQIA2S+ people is that their sexual and gender identities aren't taken seriously, believed, and dismissed as unreal, or erased (e.g., queer people told their sexual orientation is just a phase, trans people's gender identity being seen as fraudulent). The last thing this community needs is to have its experiences of psychedelics ignored, seen as "lesser than," delegitimized, silenced, or expunged if they do not fit the oneness narrative.

We are particularly concerned about this because we think the experience of oneness may be related to privilege. Our own experiences, our conversations with LGBTQIA2S+ clients and colleagues, and the growing accounts of queer and trans people's psychedelic experiences suggest that the revelation that "we are all one" may not be the dominant or most salient psychedelic narrative or experience among the LGBTQIA2S+ community. The evidence is anecdotal, and more research is needed to ascertain this. We would like to see a research study that rigorously investigates whether there is any correlation between the incidence of oneness experiences and a patient's or user's identities. Are White, cisgender, heterosexual men who developed and propagated the discourse of oneness—and are often its most vocal proponents—more likely to have that experience than queer and trans people?

One of the hallmarks of privilege is that it is invisible to those who have it unless they engage in the sustained, introspective work of becoming aware of it. In other words, people with privileged identities are less likely to think about their identity. Those with marginalized identities, on the other hand, are often highly aware of how they differ from what is considered the "norm." For queer and trans folks, healing from homophobia and transphobia often involves accepting that difference and finding celebration and empowerment in it. Given this, an experience of oneness as sameness may not only be less available to LGBTQIA2S+ folks, but being encouraged to pursue it may also be harmful. It is particularly important to bear this in mind given psychedelic-assisted psychotherapy's history of conversion therapy (most famously by Timothy Leary and Stanislav Grof). Therapeutic use of psychedelics has certainly come a long way since then, but there may be a need to reassure the LGBTQIA2S+ community that psychedelic-assisted therapy is not attempting to change us in any way. It's important to ensure protocols support and affirm the expression, discovery, and embrace of non-normative gender identities, expressions, and sexual desires. We are well on the way to this with the revision of protocols, perhaps, most saliently embodied by the removal of MAPS' requirement that the MDMA-assisted psychotherapy therapy team consist of a (cis)man and (cis)woman (one of the original reasons for having a cisman and ciswoman co-therapy team was to encourage healing mother-father experiences, which in itself is rife with cisheteronromative assumptions).

There is a growing and exciting body of research investigating the potential of psychedelics to help LGBTQIA2S+ individuals embrace their identities and heal from homophobic and transphobic trauma. Reports from queer and trans patients and users strongly suggest this may be the case. We would also like to see a rigorous study interviewing LGBTQIA2S+ psychedelic users about their experiences to see if there is any kind of meaningful and distinct commonality.

THE NARRATIVE OF ONENESS AS SAMENESS MAY BE INEXTRICABLY LINKED TO CISHETEROSEXISM

We are also concerned that the dominant narrative of oneness may be an inherently cisheterosexist one. In a 1996 interview with Playboy, Timothy Leary claimed that "the LSD experience is all about ...merging, yielding, flowing, union, communion...The natural and obvious way to take LSD is with a member of the opposite sex, and an LSD session that does not involve an ultimate merging with a person of the opposite sex isn't really complete." Like the MEQ, Leary clearly expresses oneness as sameness since merging entails the dissolution of discrete entities (here, cismen and ciswomen) into one. Given Leary's conflation of oneness with cisheterosexual sex, oneness becomes intrinsically cisheteronormative, raising the question of whether its desirability is predicated on a cisheteronormative ideal.

Leary's cisheteronormative ideal of sameness is hard to dismiss as an aberration because of how closely it ties into the history of Western philosophy and, by extension, culture. Feminist theorists such Luce Irigaray, for example, have long critiqued Western culture's inability to truly think about difference because every concept is elaborated from a cisheterosexual masculine perspective. The Western cultural framework in question is also a White, colonial one. We believe decolonizing psychedelic discourse, gender, and sexuality is a cornerstone of equitable psychedelic-assisted psychotherapy.

ONENESS DOESN'T NECESSARILY ENTAIL SAMENESS

Psychedelics such as psilocybin and LSD do, indeed, affect the brain in ways that may be conducive to experiences of oneness. While some no doubt experience that as a revelation of sameness, we suggest we need more qualitative data and a more expansive vocabulary to talk about these experiences. Anecdotal evidence suggests that oneness doesn't necessarily mean sameness. It can also be an experience of interconnection that recognizes different entities and understands them to coexist in an ecosystem of interdependence. Or, it can be a powerful experience of common humanity that still recognizes distinctiveness (*both* the same *and* different). We need to get curious about what patients and users mean when they say "we're all one."

ONENESS ISN'T INCOMPATIBLE WITH IDENTITY-BASED APPROACHES TO EQUITY

But even if the realization of oneness as sameness is, in fact, the ultimate truth to be revealed by psychedelics, that truth would not be incompatible with identity-based approaches to equity. The logic in rebuttals to equity efforts is that focusing on difference is antithetical to the higher truth that "we are all one." Such arguments want the functioning of the psychedelic community to reflect that which the medicines teach us, but identity politics are not, in fact, at odds with oneness.

The realization of oneness can be seen as a recognition that identities are socially constructed. They have no biological or ontological reality (we are the same at the core), but they come into being through networks of power (medical discourse, legal apparatus, cultural norms, etc.). Thus, for example, as Michel Foucault highlights, heterosexuality and homosexuality are nineteenth-century inventions. People were engaging in "same-sex" acts before then, but those acts were not understood to say anything about the nature of those engaged in them.

However, as feminist and queer theorists have long been arguing, just because an identity is socially constructed doesn't mean that it does not have tangible effects or that the recognition of its fictiveness is enough to dismantle it. As Christine Delphy puts it, social construction is not just social conditioning or socialization that could be transcended through a shift in consciousness. It entails the whole power of society, including its social practices, institutions, etc. There is a material reality to these socially constructed identities that cannot be willed away but needs to be named and addressed in order to be dismantled. Queer theory, which criticizes identity politics because of the normative power of identities, also recognizes the need to strategically organize around them to change the material circumstances underpinning them. To put it another way, the realization of a post-identity world requires attending to the material conditions that produce identities. Because psychedelics can show us that identities are socially constructed, they can call us into, rather than out of, working for equity.

SECTION FIVE

Sex and Power

Dating My Ayahuasca Shaman: Sex, Power, and Consent

"MARY"

I ENGAGED IN WHAT I THOUGHT WAS A CONSENSUAL RELATIONSHIP WITH my shaman. After many years, I have realized that what I really needed was for him to hold space for me, not have sex with me. But, at the time, the interaction was confusing, especially since I had just ingested a powerful hallucinogenic. Following an intense week of ceremonies at a private retreat center, my shaman approached me to offer connection and physical touch. I consented. I felt moved, even flattered. I was sitting near the firepit trying to warm up. The retreat had taken place in winter months, and I was exhausted and cold after the evening's ceremony. I wasn't ready to settle into bed in the dormitory space yet because I was still having intense visuals, so I situated myself in front of theheat to ground myself. My shaman appeared with the offer of a blanket, which felt like a warm, welcomed gesture. The gesture escalated quickly, from a blanket to a hug, from a hug, to a kiss. And, because my nervous system was so activated and my senses were heightened, all of this felt good. My judgment, on the other hand, was still impaired.

Though I didn't know anyone at the retreat center, I went there based on the recommendation of a trusted female friend. This, paired with my willingness to put my faith in the man who was guiding the ceremony, seemed enough at the time to allow me to trust my own actions. After all, I trusted him earlier in the evening when he helped me by singing me through a difficult and embarrassing purge. So, when it came time to trust his gestures afterwards, I did. The intimacy seemed consensual. But what I've realized is that, because of a shaman's perceived expertise, a participant often places a great deal of trust in them to hold a safe space. A shaman's responsibility is to maintain professional boundaries and understand the sensitivities that come alongside participants' experiences in ceremony space, and this safekeeping can be violated if the healer-participant power dynamic is exploited.

POWER AND RESPONSIBILITY

I was not completely unaware of these power dynamics, and I had a mixture of feelings following the initial physical interactions we had, despite how good they felt. Afterward, I pushed explicit dialogue with him regarding consent, power dynamics, and being under the influence of a powerful psychedelic. But the intimacy that followed my experience had already imprinted on me, and it seemed impossible to unravel myself from him. We continued to connect. The connection evolved into a relationship. In retrospect, I see that the initial ease of these conversations helped bolster my denial that the developing scenario was less than ideal. As a practicing shaman, it was his responsibility to initiate these discussions with me. Ideally, these conversations would have taken place when we were both sober, rather than post-ceremony, when ceremony participants often feel open and are still processing their experience.

It is not uncommon for people to experience increased libido or sexual thoughts during an ayahuasca ceremony. However, communities where ayahuasca is used often adhere to guidelines recommending decreasing sexual activity pre- and post-ceremony as part of a *dieta*. In the hours before I was approached with an offer of physical contact by my shaman, I was experiencing wildly visceral sexual hallucinations. My libido was heightened, and this was not a clean and integrous lead-in to physical contact with the person who was facilitating my experience. Ideally, he would have anticipated and recognized the potential for my increased libido during the ceremony, and left me well alone.

A WIDESPREAD PATTERN

After my relationship with him, I learned that he repeated this healer-participant relationship pattern. Although I maintain that our relationship was by-and-large consensual, the continued pattern of his relationships seemed worrisome to me. In the years following, I see that I had been vulnerable and open to suggestion in those initial moments when the connection was being formed, and I wondered if the other women experienced this with him as well. I felt similar to Emily Sinclair when she stated, "I did not realize at the time that this behavior was widespread but, instead, blamed myself for being naïve."[1]

In my Western community, I have witnessed a handful of shamans act outside of propriety. I have listened to firsthand accounts from both female and male survivors who have been harmed by sexual abuse in these contexts. One shaman slept in a bed next to a participant after a ceremony to help calm their nerves. One shaman bragged that a new participant had the "hots" for him. One shaman asked a participant for a date after the participant shared to the group in the integration circle the deep realization of longing for a life partner. One shaman "healed" participants' sexual trauma by having sexual contact with them. One shaman used spiritual bypassing in hopes that his history of sexual assault would be dismissed, ignored, or dealt with outside of legal channels.

SEXUAL ASSAULT AND SPIRITUAL BYPASSING

The New Age movement has a tendency to sanitize the experience of ayahuasca, propagating an overreliance on a "love and light" mentality that can lead to spiritual bypassing. This can deflect responsibility for the accused, resulting in survivors finding it difficult to make strides toward criminal investigation. It can also cause challenges as survivors try to heal and make sense of their experience in a grounded way. The last-mentioned individual in question in the above paragraph responded to a public call-out by requesting community rather than legal counsel, referencing bestselling author Don Miguel Ruiz Jr.'s book based on Indigenous Toltec wisdom, *The Four Agreements*, one of which is, "Don't make assumptions." Later, he quoted abridged language of an Indigenous Hawaiian tactic of forgiveness and understanding, *Ho'oponopono:* "I love you, I'm sorry, please forgive me, thank you." The use of this borrowed language in a context in which the use of Indigenous plant medicine has already been perverted to sexually abuse participants is particularly egregious.

There is less of an emphasis on this love and light mentality in the Amazon, as the darker sides of this multifaceted, powerful medicine are acknowledged and well-known. This is not to say that Indigenous communities do not face complaints of sexual assault—they certainly do. All shamans are human beings, whether they are practicing in private ceremonies,

ceremonies sanctioned by religious institutions, or within Indigenous communities with long histories of ayahuasca use, human fault is possible. In fact, there is a danger to thinking that only so-called "inauthentic" shamans may commit abuses, as it feeds into a tendency to exoticize and idealize Indigenous communities. Daniela Peluso points out that Western shamans who have committed assault are commonly called "inauthentic" and "not real shamans" by bloggers, but this sort of language idealizes and "romanticizes the 'other'."[2]

WORKING TOWARD ACCOUNTABILITY

Authenticity is irrelevant. Instead, possible explanations for the widespread increase in cases of sexual assault include both the need for secretive use of the medicine and the lack of an official organizational structure, making it more challenging to hold perpetrators accountable in sexual assault cases. Many of the cases mentioned above took place in small, private ceremonies held in places where ayahuasca use is illegal, which presents a challenge when it comes to accountability. Although such ceremonies have their own guidelines, ethos, and rituals, there are benefits to religious models; for example, the Santo Daime church, which has its own social mores that reinforce a shared set of values and measures for community safety. Such models based on religious doctrines have the ability to provide some harm-reduction strategies, however, it is also widely known and acknowledged that Santo Daime and other religious communities are not immune to instances of sexual assault. In a community for which use of this medicine is still largely unchecked and unregulated, organizing toward accountability and a strong, shared ethos is critical.

The Chacruna Institute is one organization that strives to bring awareness to cases of sexual abuse. Chacruna's "Ayahuasca Community Guide for the Awareness of Sexual Abuse"[3] outlines an array of common scenarios that may help ayahuasca communities feel more aware, resourced, and informed. And, in a companion to this resource, the Chacruna Institute details the current legal situations in a handful of South and Central American countries where ayahuasca is used.[4] As we move forward into a future that will likely see an increase in the global interest and demand for ayahuasca, spreading awareness about safety precautions and the realities of sexual abuse in communities both Indigenous and Western will continue to be vital.

It took a great deal of time for my understanding of my relationship with a shaman to take the shape it has today. And, although I hold no grudge, I see now that the nurturing, connection, and intimacy that took place in the context of a sexual relationship would have been healthier, more honest, and more healing within a secure therapist-patient connection. My hope is that we, as a community, can learn the importance of maintaining such ethical dynamics, so that harms are minimized, healing is emphasized, and the patterns that were enacted in my relationship will not be repeated elsewhere.

Sexual Assault and Gender Politics in Ayahuasca Traditions: A View from Brazil

GRETEL ECHAZÚ, PHD AND PIETRO BENEDITO, PHD

IN AUGUST OF THIS YEAR, A BOMB EXPLODED IN THE WORLD OF NEW AGE practitioners: The Brazilian guru, Prem Baba, a charismatic religious leader who propelled a New Age movement with a worldwide following, was accused of sexual abuse by two women who for years had been a part of his closest circle of adepts. Janderson Fernandes (Prem Baba's birth name) was exposed by his victims to a major Brazilian news source, the Época Magazine, as having used his charismatic leadership to rope them into a series of sexual activities, which he called a special kind of "tantric treatment," performed throughout by him and each woman "consensually."

One of the women opened up after finding herself experiencing serious panic attacks. After the news bomb hit, Prem Baba publicly recognized his "mistakes," minimizing them as "part of his learning path," and characterizing the denunciation as a kind of "dark force growing in the hearts of his people." The women's accusations were depicted on a YouTube post by Janderson as a flaw of their spirituality, rooted in inexplicable greed and pain.[1] Afterwards, many of his closest followers publicly rejected his version of events and stood up for the two women, sparking a massive exodus from his movement.

AYAHUASCA CIRCLES IN BRAZIL

In Brazil, legal regulations determine that one can engage in an ayahuasca experience only in religious settings, of which there are three main ones: Indigenous traditions, the "ayahuasca religions" (Santo Daime, União do Vegetal, and Barquinha), and the increasingly popular neo-ayahuasquero centers. Most people drink ayahuasca within the religious ayahuasca groups.

The ayahuasca churches have been present in Brazil since the 1930s. They combine diverse cultural influences, such as Indigenous, European, and African-Brazilian through the framework of popular Latin American

Christianity. Neo-ayahuasquero centers are characterized by the more fluid and less institutionalized use of ayahuasca and New Age religiosity. Janderson Fernandes began his career as a spiritual leader in one of these nuclei when, as a psychotherapist, he started sharing ayahuasca with his followers.

These non-Indigenous ayahuasca circles in Brazil are demographically predominantly middle-class, White people of often surprisingly conservative views. In an elitist and conservative exception to the popular paradigm of the psychedelic revolution, the religious and therapeutic experience of ayahuasca in urban contexts is available mainly for those who can afford or have access to it.

DESCRIBING HOW THE FAMILY IS THE UNIT OF SOCIETY AND HOMOSEXUALS WERE OUT OF THE "CORRECT" ORDER OF THINGS

An example of this is the alignment of some União do Vegetal (UDV) leaders with the reactionary campaign of the recently elected presidential candidate, Jair Bolsonaro, who harbors extreme right-wing views. In September 2018, the ex-general leader of UDV in Brazil, Raimundo Monteiro de Souza, recorded a WhatsApp audio calling for followers to support the fascist candidate and stop a supposed "moral crisis" pervading the country.[2] This did not come as a surprise, as the UDV had in the past expressed reactionary views on multiple fronts. One clear example is an internal UDV statement, which was later leaked on the internet, describing how the family is the unit of society, and homosexuals were out of the correct order of things. Also, it is important to remember that, within this group, women cannot reach the top of the hierarchy as masters, a position restricted to men, a rule that has been rarely defied.

ALARM SIGNALS

As ayahuasca grew more widely popular in Brazil, cases of sexual assault and misconduct toward female participants in ritual settings began to emerge. The exact number of cases is unknown, since many of them don't come to light. Fortunately, this issue is increasingly illuminated by participants, academics, and activists in a growing international movement that denounces practices previously silenced within ayahuasca circles.

In this article, we want to bring out the strategic relevance of a gendered perspective on religious groups in contemporary contexts, making visible the scenarios that make possible abuses of all kinds, especially those of a sexual nature.

TOOLS FOR ANALYSIS

The complexity of abuse and, specifically, of sexual assault in religious circles has been studied under a number of perspectives,[3] but is certainly more complex in cases where a psychoactive substance is included at the core of the religious practice. Such is the case in the evaluation of sexual assault cases in the ayahuasca religious context. What is to blame for the violations: the psychoactive plants or the sexist hierarchies prevalent in those contexts?

Sexual abuse in ayahuasca circles is a very delicate matter that involves, first, a critical reflection about consent and autonomy in the context of hierarchical relationships and, second, thoughts about the influence of ayahuasca in this dynamic. There is a tension between the human rights of collectives, such as the ayahuasca religious groups, and the human rights of individual subjects, particularly those of women, LGBTQ+ people, disabled people, and others in situations of social vulnerability respective to their perpetrator.

Our proposal is that, in Brazil, some religious groups are based in a hierarchical and sexist structure in which women are "essentially" different from men. This affects women and LGBTQ+ people in different ways. There are some ayahuasca religious institutions and neo-ayahuasquero groups in Brazil that tend to perpetuate this difference regarding their choice of leaders, the configuration of rituals, and the establishment of genealogies of their divine masters.

Following cosmologies and specific liturgies, it is common among some ayahuasca religious institutions to find a strong sense of hierarchy that empowers commanders. (Of course, there is much variation between different churches and organizations and we are using generalizations here for the sake of our argument.)

By presenting their leadership as destined to a spiritual mission, religious authorities often make it difficult for their words and actions to be questioned by followers. Also, the divisions between men and women are strongly based on traditional Brazilian gender roles. Often, men are defined as active rulers and women are portrayed as docile and natural followers.

In some neo-ayahuasquero groups, even though they generally seek to differentiate themselves from institutionalized religions, and embrace a more fluid structure, a good number of leaders also tend to assume hierarchical positions.

This situation within these religious communities is at risk of growing more rigid in the face of the rising political wave of fascism—as epitomized in the recent election of far-right candidate Jair Bolsonaro to the Brazilian presidency. Between this rigid internal sexist hierarchy and an increasingly conservative and hostile political environment, we anticipate the growth of sexual assault and other abuses within ayahuasca groups in Brazil.

The Brazilian State favors freedom of expression and inner organization within the ayahuasca religions. But, like any organization subject to legal scrutiny, its principles should not compromise human dignity, and, despite the importance of thinking the human rights of collectivities the dimension of the individual remains fundamental.

Human dignity and human autonomy are values that should be translated into dignity and autonomy for every subject. No religion should impose a perspective obliterating the fundamental human rights of its participants. In the case of women, LGBTQ+, and other groups in similar situations of vulnerability, this attention should be redoubled.

PROPOSALS

To recognize, reflect about, and change this reality, we propose a few actions towards the politicization of ayahuasca circles in Brazil. First, we support questioning the cultural autonomy of practices that take place in ayahuasca religious rituals, intertwining those experiences with gender, race, ethnicity, class, and other concepts that are sensitive to social differences within settings. In this sense, we argue that abuse is not linked to a unique practice, but to many. For instance, racism and classism might manifest alongside the expression of sexual abuse at times.

On a more concrete basis, we call for the production of materials discussing assault and consent to ayahuasca groups worldwide, highlighting what is and is not a healing practice in a pre- and post-ritual context. We support the production and sharing of such resources between heterosexual women, bisexual, gay, and trans people attending ayahuasca rituals in religious contexts as a means to prevention. We also recommend the creation of a headquarters for receiving complaints, offering a space to listen and to assess experiences of abuse in these contexts as a means of treatment, as well as multiplying platforms for complaint and redress that are accessible to civil society.

And last, but not least, we wish to highlight the importance of linking ayahuasca's role as a companion plant to social and community development in local, integral, holistic ways to our unfolding story as species who met, built together practices with a purpose, and helped remake society in this era of the Anthropocene.

Abuses and Lack of Safety in the Ibogaine Community

JULIANA MULLIGAN, BA

JUST A DASH OF DEATH

My first encounter with the ibogaine community nearly killed me, but it also saved my life. This exemplifies the juxtaposition inherent in ibogaine community dynamics. Ibogaine offers a powerful experience that attracts those seeking miracles, but it also attracts those seeking to be the beholders of miracles. Many working in the ibogaine world have spent long years suffering through destructive behaviors while also searching for purpose and meaning—what better path to finally harness for your life's work than the one that changed your own life? But the line between providing healing to others and holding tyrannical power over others is easily blurred in a community where one might almost conclude that mental instability and unaddressed trauma are required qualifications for the job.

I am one of the many people who became an evangelical ibogaine advocate and aspiring provider. After seven years of opioid dependency, and many other years suffering with depression and anxiety, ibogaine treatment was pretty much the only treatment I hadn't tried. Although my treatment was a success, in that I left opioids behind for good, I also ended up in an ICU in Guatemala City for ten days on an external pacemaker after experiencing six cardiac arrests. The guys providing my treatment were not following proper safety protocols, and this isn't a unique occurrence in the world of ibogaine clinics and ceremonies, most of which operate out of the bounds of any governing body. Regardless, the benefits I experienced from the medicine were so profound that I knew I wanted to work with this substance. Discovering that your life of disaster and suffering is suddenly transformed into a kind of priceless training or internship (as I like to refer to my opioid using years) is quite the golden discovery. Working in ibogaine as a former substance-dependent person offers this deep sense of purpose and meaning.

As I was introduced to the community of ibogaine providers, researchers, and doctors in 2012 at my first GITA (Global Ibogaine Therapy Alliance) conference in Vancouver, the reasons behind my hospitalization in Guatemala became obvious. The cardiotoxic effects of ibogaine mean that pre-treatment screening, intensive medical monitoring, and diligent adherence to the clinical safety protocols is non-negotiable. The clinic I was treated at in Guatemala made a number of mistakes that became crystal clear as I watched presentations from Dr. Ken Alper, a psychiatrist and researcher at NYU, and others. But research, case reports, and science alone aren't enough to stop dangerous practices in a community that operates outside of borders, the law, and any kind of supervision process.

INSTA-SHAMAN

Although many make claims that ibogaine is like "10 years of therapy in one night," it cannot instantly erase a lifetime of trauma and unhealthy patterns, and it certainly does not "cure" serious mental health issues. Don't get me wrong, ibogaine is an amazingly powerful medicine, and none of the other treatments I tried for my opioid dependency came anywhere close. The so-called "resetting mechanism" of ibogaine isn't completely understood scientifically, but there is no denying it once you have experienced it. Waking up in the ICU in Guatemala, I literally felt like I had shed my old self and stepped into a new life altogether. I found my mission in life, was excited and inspired about my new path, and seemingly, like magic, I skipped the crushing fentanyl withdrawal I had coming to me. But, while I left my old destructive relationship with substances behind, the experiences from my childhood still dictated my emotional reactions. The unraveling of lifelong patterns and behaviors was only beginning for me. Ibogaine is an amazing tool, but it's merely a door opener.

The following days, weeks, and even months after stepping out of a drug dependency with the help of ibogaine can be a magical time when you experience a connection with life that potentially hasn't been felt since childhood. It's so magical that many decide that the right thing to do is to open up an ibogaine clinic immediately, or to go work at one that already exists, without taking time to do internal work or just to learn how to live without chaos. Suddenly, you're on a God-given mission to help others. You might even imagine that you have a new psychic ability to "save" others.

Quite abruptly, you have gone from longing and searching for something that's missing in your own life to believing you're the person who is going to deliver people from their own arduous longing for purpose.

The issue latent within this, is that weeks or even months is not sufficient time to process the internalized abusive patterns that have been absorbed from the patriarchal, colonizing, and White Supremacist societies that many of us are raised in. Not only is it not enough time, but it's probably not going to happen at all without a serious willingness to self-reflect and without the right therapeutic support. The explosive enthusiasm to work in ibogaine isn't just about finding "your calling," it's also born out of an anxiety-driven sense of urgency to find the next external thing that will make your life complete. This is one of the ultimate curses of modern society: We are programmed to believe that material items, other people, drugs, or a certain career is what makes us complete and whole. This desperate seeking, and the clinging onto ideals despite clear warning signs, does not go away with ibogaine. In fact, without the right guidance and support, ibogaine can make it worse. In other words, ibogaine can cause ego-potentiation.

The "shaman complex" is a good way to sum this up. What better way to fix years of feeling like the scum of the Earth than to suddenly become a person that is praised and looked up to and helps others? Stepping into the role of "healer" is perfect for anyone with sociopathic or personality disorder traits. You get to have power of others, receive admiration, potentially claim that you're some kind of prodigy, and actually make money from all of this. In this way, an unregulated and unmonitored community can become the perfect grounds for dangerous and unstable people to thrive.

I'm not saying there aren't talented, stable, and loving people working in the community—there definitely are—but, from what I've witnessed, this is not the vast majority and a good portion of the safe practitioners are in fact not men. It's not cut and dry by gender, and there are unsafe women perpetuating abuses too, but the scariest behaviors I see are perpetrated by the men in the community. Although women suffer from patriarchal societal practices, men are also negatively impacted by them. From a young age, men are taught to hide their true emotions, maintain an appearance of toughness, get recognition at all costs, be a leader, and that these things define who you are. From what I've witnessed, it seems that this conditioning can often lead to a terrifying display of power seeking, inability to take responsibility, and self-righteous vigor, all disguised as spirituality and service. Internalized patriarchy can also lead other genders to exhibit these qualities as well, however, because the voices that receive platform in this community have mostly been male, these dangerous behaviors have been most apparent among the men.

WOUNDED HEALERS, WOUNDED DEALINGS

At the first clinic I trained at, each week, the owner left four to six people (usually predominantly men) under the care of one female nurse for the night of their ibogaine flood, with no doctor or anyone else on site. When I and others in the community raised concerns after witnessing a seizure there, the reaction of the owner was of defensiveness and arrogance. He claimed to have never had a medical emergency and that one would not happen at his clinic, ever. A few years later, a patient died, and many long-time members of the community were not surprised.

Unfortunately, that story is nowhere close to the worst. I have heard multiple accounts of people being left in hotel rooms alone during a flood dose, sometimes, the provider never returns. In one of these stories, a woman woke up in a hospital in Thailand with the provider nowhere to be found, having no idea what happened, and her money gone. There are also many accounts of outright sexual abuse, mostly perpetrated by men upon female clients. A large proportion of clinics are run by men and with a male majority staff. More often than not, women and non-binary folks who come for ibogaine treatment have a history of sexual violence or patriarchy-related trauma. In order to responsibly work with women and other marginalized groups, this would require specialized trauma-based training with a focus on sexual and systemic racial trauma. But this isn't happening, and there is no person or organization able to enforce responsible practice requirements or hold people accountable.

Beyond neglectful or unsafe clinical practice and outright physical or sexual abuse, there is also harder to distinguish psychological and emotional abuse. It can be incredibly hard to identify and take action against this type of abuse, due to the manipulation the perpetrators are often so adept in. The other issue is that it's difficult to get people to rally in support around victims of this type of abuse because the perpetrators are so skilled at presenting themselves in different ways to different people. Some are highly regarded in the community but are carefully hiding violent behaviors.

There is also the problem of stigmatization and ostracization between community members. You would think that working in a healing practice would foster only compassion, understanding, and forgiveness, but, unfortunately, the volatile expression of unaddressed trauma carries into the land of ibogaine too. A few years ago, a friend working in Mexico had a relapse which led to some issues in the community. Instead of a loving intervention, other providers in the nearby community cut off communication and stigmatized him. He has since left the ibogaine world heartbroken, discouraged, and disconnected from the people he had once thought were his family. This story has happened so many times that even in the midst of a dangerous relapse, when life is on the line, people are too petrified of judgment and ostracism to reach out for support in their own community.

IBOGA CONSERVATION

Abuse does not just work from person to person, but can be from person to plant, and culture as well. According to the conservation and sustainability organization Blessings of the Forest (BOTF), the rapid popularization of iboga in the West has led to plants being stolen out of protected forests in Gabon, leaving traditional Bwiti practitioners no longer able to access their own medicine.[1] There are constantly new iboga sellers popping up online, with no official way to verify that their source is legal or sustainable. There is also the added complication that iboga is not usable as medicine until 5–10 years of growth, so, replenishing plants is no simple task. The best way to guarantee you aren't using stolen iboga is to use ibogaine converted from voacangine, which is derived from the voacanga tree. This tree grows quickly and plentifully all over West Africa, so it's a much more sustainable source of medicine. But many practitioners insist on using iboga-derived medicine, often citing that the "spirit of the plant" is missing in voacanga derived ibogaine. I and others have personally witnessed the same transformational experiences with ibogaine derived from voacanga. In my opinion, the "spirit of iboga" is not limited in its ability to show up when it's called, no matter what plant is used. Applying science-based logic to plant intelligence and spirituality will give us a very narrow and incorrect view of how iboga works. Besides, I imagine the spirit of iboga would greatly appreciate us giving her some time to rest and replenish. Neglecting sustainable sourcing of medicine is a continuation of colonization. At the very least, clinics should be donating back to Gabon, and BOTF, to ensure iboga doesn't cross into extinction and to safeguard future access to iboga for the traditional peoples who use it as a part of their religion.

ACCOUNTABILITY, SAFETY, AND THE FUTURE OF IBOGAINE

Legal regulation, the medical model, and FDA approval are not necessarily the answer for safe psychedelic use. Although aspects of those systems might help in certain situations, there is also reason to believe that this places power in the hands of oppressive governments, a medical system that is often violent and stigmatizing to marginalized populations, and pharmaceutical companies who prioritize profit over people. You might ask: How then can we ensure safe practices and accountability in the ibogaine community?

When GITA was still functional, the publishing of the clinical guidelines in 2015 was a major leap forward. Now, there is a solid, well-researched, and reviewed publication outlining how to facilitate a safe ibogaine treatment. The problem is, there is no way to enforce these protocols and no way to hold practitioners accountable when they aren't followed. For the most part, clinics operate in places where a client can die and no legal consequences will occur. For example, there was a provider working in Costa Rica who had multiple deaths across multiple Latin American countries and was never held accountable. GITA once offered a grievance process for people needing support around bad clinic experiences, but ways to address the problems with the provider were limited. They could simply choose not to engage in an accountability process.

So, what *are* the ways to build accountability, responsibility, and better support into the ibogaine world? I am currently a part of the cofounding of a women and genderqueer-led collective for people working in the community called The Root Ibogaine Collective. We are a group formed to promote equity, ethics, safe practices, sustainability, and accountability in the ibogaine community. One idea in development is to offer a training and certification process to clinics that would address clinical safety, but also offer trauma-informed training. One of the requirements for our approval would be that each employee of the clinic be involved in a personal therapeutic process and that a certain percentage of the staff be women. The hope is that, gradually, enough clinics would want to engage in our certification, so that it becomes something that people look for when selecting a clinic. We are also planning to offer weekly meetings and check-ins for providers in order to create a better support system for them.

PSYCHEDELICS AND RESTORATIVE JUSTICE

Something else that is much needed, not only in the ibogaine world, but also in the greater psychedelic world in general, is a well-constructed space for accountability and restorative justice processes. So often, the way that harm is dealt with is either using victim blaming and shaming, or the call-out and cancel approach. Both things are rooted in abusive, patriarchal, and colonizing methods of violence. There is a time and place for people to be blacklisted, but more often, there are other routes where healing and education can occur. Deaths, exploitation, and trauma will continue to be perpetuated in our community if we don't start to set up compassion-driven structures to address harm.

Of course, healing communities attract wounded people, and we all still have work to do on ourselves, and that's okay. But, if a person is doing harm, and continually refuses to be responsible for it, then a community-led intervention is needed. The hyper-focus of many on getting FDA approval for ibogaine is dangerous, because that will not address the systemic and interpersonal abuses that continue to happen. Before we can spread ibogaine to a larger audience, we must address the dangerous issues we have now.

The abuses in the ibogaine community are not just limited to the provider-to-client relationship, but also occur between those who work in the community. One of my main inspirations to write and work on these issues comes from something I recently experienced personally. A person who I collaborated with and was quite close to in the community for many years became very emotionally violent towards me last year. As I watch him receive accolades and respect for his work, he still won't accept responsibility for this behavior, which is very concerning and painful to witness. We can't discuss systemic violence without discussing interpersonal violence, because what happens in our close relationships creates the materials that build the larger system. This experience has made it clear to me that new systems need to be in place in the community to support victims of abuse. I would like to call on anyone experiencing harm in the community to speak up about their experiences and ask for support. It's takes a collaborative community approach and a willingness to be vulnerable to create healing structures that address harm.

Psychedelic Masculinities: Reflections on Power, Violence, and Privilege

GABRIEL AMEZCUA, MA (C)

THERE IS NO DOUBT THAT THE GROWING INTEREST IN THE USE OF PSYCHEdelic substances has triggered a wave of increasingly critical attitudes toward the neoliberal way of life and towards the social systems that shelter it, where the patriarchy is of major concern. Characterized by attitudes of domination, extractivism, and competition, the negative aspects of the patriarchy as a social system become more evident from the perspective of a person under the extreme sensitivity of psychedelics. No wonder, then, that many of the groups of the so-called "new masculinities" that I have met are made up of a majority of people who use or have used these substances. I do not think it's naive to say that psychedelic substances have helped to build masculinities that are more sensitive to our environment and more willing to empathize with others.

However, in spite of this, the psychedelic community is plagued by scandals that involve physical and mental abuse perpetrated by male facilitators and researchers. Far from being more sensitive or empathetic, some men even seem to build up notions of self-pride that foster sexualizing and abusive versions of themselves.[1] For this reason, talking about masculinity and patriarchal behaviors within the psychedelic movement has become more necessary than ever before, as is talking about possible strategies to control their damage through self-critical and constructive approaches to masculinity.

As a response to what some may call a global masculinity crisis, it has become increasingly common in recent years to hear about movements of new masculinities, also called by some "healthy masculinities" or "conscious masculinities." These groups seek to become a movement of non-hegemonic men developing in opposition to toxic, dominant, and violent masculinities. Although its existence goes back to the egalitarian sociology of the 60s and 70s, today there are multiple approaches and perspectives.

Overall, these groups lack cohesion and represent a vast terrain of contradicting opinions that have been unable to generate collective organization or sufficient force to transform a common reality. Nevertheless, at this point in time, these budding ideas also provide different potential avenues along which the movement can ultimately walk.

As an anthropologist, I have found in the groups of masculinities both ethnographic research material as well as a place of learning and reflecting about what it means to be (or not to be) a man. My first group of masculinities emerged in Mexico in an environment of human rights professionals and activists orientated towards issues of sexuality and gender equality. Our interest focused mainly on self-criticism and deconstruction of violent internalized attitudes. It was painful and often uncomfortable, but we grew through understanding the unconscious ways in which we exercise patriarchy in our daily lives. Our groups were built on dynamics of sharing and mediation circles, expressing our pain, admitting our frustrations, and denouncing our own sexualizing or violent behaviors, seeking among our peers advice that could help us heal and improve our practice of masculinity.

In Mexico, I also became familiar with the global organization Men Engage. This is a network of masculinities whose main objective is to speak out against the way in which men's roles are scripted with violence. It takes a critical look at how violence is expected from men, while they are shunned for it at the same time, and how this affects every aspect of their lives, from interpersonal relationships to politics. When I moved to Europe and I was invited to other new masculinity groups, I was excited and motivated to see that the movement has gone global and was changing men all over the world. Yet, I was certainly confused by the stark difference that I experienced between the perspectives taken. I do see the value of the groups that I have encountered so far, but I think that the understanding of the conflict, and thus the direction taken, stem from very distinct cultural approaches.

These new experiences are mostly composed of middle-class, White, Western men, coming together to pay homage to "sacred masculinity," talking about the right to be vulnerable, and to remember that masculinity can be beautiful and emotional. Retreats with catchy names such as "Rebel Wisdom," "Awakening Masculine," or "The New Masculine Program," are full of ayurvedic food, sexuality enhancing techniques, and coexistence dynamics that often point to losing the fear of being vulnerable among other men. In the end, it is true that one of the problems that we often experience within masculinity is our difficulty in sharing our feelings. However, I think it is naive to believe that this is the essential problem of masculinity.

Before going forward, I would like to note that I do not want to delegitimize these ideas. I really believe that it is important to talk about the masculine capacity to be vulnerable and create safe spaces where we can be sympathetic to each other and share our emotions. However, the deconstruction of patriarchy requires much more than building an altar for emotional and empathetic men. The work of new masculinities requires a sincere confrontation with the power imbalances between men and women, without feeling uncomfortable to explore the fact that there are places within us from which we exert power without being fully conscious of it. It requires us to be honest about how we exercise control and privilege through being men and, above all, it requires a work of deconstruction of the patriarchal attitudes that we take for granted in our daily life: extraction, projection, competitiveness, colonization, domination, etc.

As in the psychedelic movement, the new masculinity groups—and even more, the groups of the so-called "feminist men"—are riddled with scandals of sexual abuse and inappropriate behaviors towards women. Clearly, belonging to a group of new masculinities does not guarantee an immediate evolution towards better versions of ourselves. I think it's important to highlight this. Likewise, I think it is important to stop clinging to sensitivities around concepts such as "toxic masculinities." In the end, it is not unreasonable to say that patriarchy is toxic, Western colonialism is toxic, and capitalist consumerism is toxic. These are the manifestations of compulsions and appetites that—as a society—we must learn to recognize, limit, and eventually dissolve. In the same way, masculinities tend to toxic behaviors that we as men have to learn to recognize, limit and dissolve, with the aim of balancing the power between men and women, as well as eliminating the intrinsic violence of war-like masculinity.

Groups of new masculinities that focus on generating a better self-perception of masculinity and fraternal congeniality among men seem useful to me as self-help groups, but not necessarily effective in the creation of socially aware and responsible masculinities. It's important to point out the differences in approaches between the new masculinities whose objective is gender equality, including the deconstruction of patriarchy, compared to the new masculinities that seek to redefine masculine identity through sensitivity and the loss of anger. While the first approach explores the privileges and burden that arise from being a man in society, the other seeks to improve the capacity of emotional expression and the way of relating among men through quasi-spiritual therapeutic approaches. Both of these approaches have valid objectives. Nevertheless, only the first approach actually allows for open self-criticism because of its focus on the analysis of the issues that revolve around masculinity. Moreover, it is relatively common for this approach to hurt "male sensibilities," precisely because it is not easy to admit the problems that we are part of when we exercise our privileges.

The lessons learned in the new masculinities movements have much in common with the lessons we must learn in the psychedelic community. If we have grasped anything in recent years, it is that, although the psychedelic experience has an immense transforming potential capable of producing experiences of spiritual emergence and greater empowerment, this is, however, not immediately synonymous with being "better human beings." As mentioned before, there are already too many cases of self-proclaimed facilitators and shamans who, regardless of the number of accumulated psychedelic experiences, persist in behaviors of abuse, arrogance, and domination, that is, primordially patriarchal behaviors.

Regardless of the intensity of the spiritual emergency, psychedelics do not guarantee critical reflection. Although they definitely facilitate opening the door to self-reflection, without seriously working on analyzing our compulsions, our fears, and our anger—the psychedelic experience runs the risk of remaining a mere aesthetic appreciation of altered states of consciousness, without necessarily changing the way we exercise that consciousness. Being part of groups that highlight the spiritual experience and facilitate human expression can be used as a green-washing of the conscious kind, or what the author and academic Monica Emerich calls a "spiritual-washing," through which people assume a personal evolution only because they participate in dynamics or buy products that have the stamp of "spiritual awareness."[2] The ceremonies with sacred plants and affective group dynamics become apparent guarantors of better human quality, but with a lack of self-critical content, lasting change is unlikely.

The psychedelic masculinities, like other masculinity movements, need to maintain self-reflective and deconstructive stances on patriarchal, consumerist, extractive, colonialist, and neoliberal behaviors that have put at risk the integrity of hundreds of Indigenous cultures, the environment, and the right of women to live in an equal society. Otherwise, they simply become spaces to feel better about ourselves and our immediate reality. This reflection is not only limited to masculinities, although, in terms of privilege and violence, we have a lot to work on. My way of perceiving a truly responsible psychedelic movement is one that uses psychedelic substances for personal work in questioning our privileges and biases. If we are White, we must give space to reflect on our race privilege and meditate on how we can be part of an effective decolonization. If we have enough money, we have to allow a space for critical analysis of our class privileges and reflect on how we can help generate sustainable modes of existence. If we are men, it is our responsibility to admit our gender privilege and submit ourselves to a self-critical analysis to diminish our violent, macho, and competitive behaviors, aiming to build truly egalitarian societies.

Using psychedelic substances to heal ourselves and our society also requires our commitment to work on and change all those behaviors established when we exercise our privileges. Whether we like it or not, every process of healing and evolution brings its own sacrifices, and one of the greatest sacrifices is, without doubt, the critical, reflective, and sometimes painful analysis of those aspects of ourselves that allow us to exercise power over others.

Ayahuasca Community Guide for the Awareness of Sexual Abuse

BEATRIZ C. LABATE, PHD AND EMILY SINCLAIR, PHD

MANY INDIVIDUALS PARTICIPATE IN AYAHUASCA CEREMONIES FOR HEALing purposes, sometimes specifically to heal trauma caused by sexual abuse. Considering this, it is especially disturbing to discover that sexual abuse is also quite prevalent in ayahuasca and shamanic healing contexts. Several cases, some quite high profile, have come to light in recent years spanning different kinds of abuse across diverse ayahuasca ceremonial contexts. Yet despite sexual abuse and harassment being prevalent within ayahuasca circles, many participants seeking ayahuasca healing are still unaware of the problem and can unknowingly end up in a vulnerable situation. Indeed, one of the first obstacles we face in attempting to address sexual abuse in the ayahuasca community is the quite widespread disbelief that sexual abuse is indeed a problem. The Chacruna Institute for Psychedelic Plant Medicines believes that the more people who learn about past and potential sexual abuse, the greater the chances to combat it. Sexual assault is *always the fault of the perpetrator* and it is the responsibility of all individuals within the community to come forward and speak about this. While we have no control over the perpetrators of these acts, we hope that the experiences of others can be useful in raising awareness about the typical contexts in which past abuse has occurred.

Motivated by a desire to raise awareness and help safeguard individuals and groups in ayahuasca healing contexts, Chacruna produced the Ayahuasca Community Guide for the Awareness of Sexual Abuse, an initiative of Chacruna's Ayahuasca Community Committee.[1] We chose to craft the guidelines to focus on women, since it is mostly female participants being abused by male shamans that comprise the bulk of sexual abuse occurrences. Yet, our hope is that they are of value to all. Attempting to cover diverse social and cultural settings where ayahuasca healing takes place, the guidelines have been created through a collaborative process with many experienced individuals in a wide range of ayahuasca settings across different cultural contexts and continents. This shared process has included Indigenous as well as Western victims and survivors of abuse, ayahuasca healers and ceremonial facilitators, and anthropologists who like ourselves have conducted long term fieldwork in lowland South America and have long-standing experience with ayahuasca communities. We have also attempted for these guidelines to be useful across the spectrum of potential abuse that can occur in ayahuasca settings, including verbal persuasion, invasive touching, "consensual" sex between healer/participant, and rape.

In forming the guidelines, we began by asking, why is sexual abuse so prevalent in ayahuasca circles? Apart from acknowledging that sexual abuse is an abuse of power, which occurs broadly across diverse contexts in diverse societies, we were interested in better understanding what elements or conditions can specifically be linked to ayahuasca healing contexts. One issue is the undue romanticism that can surround expectations about ayahuasca and *ayahuasqueros*, and the assumed position of trust a healer or ceremonial facilitator inhabits in the imagination of participants.

When entering into ayahuasca healing circles one may assume and expect to be entering a safe space. One may assume or expect to be able to trust the people calling themselves healers, "shaman," leaders, and facilitators of this space. The guidelines hope to shed light on the context of typical ayahuasca scenarios, and what some of the associated assumptions and expectations surrounding these might be. For instance, the guidelines hope to de-mystify the position of the ayahuasca healer as well as to draw attention to multi-cultural issues within ayahuasca community contexts that are not immediately understood or applicable outside of such settings.

While no participant can be told what they can and cannot do with their bodily autonomy, the guidelines present a series of cultural differences that have typically created confusion, miscommunication, and conflict in ayahuasca healing settings. It is important to note that sexual abuse of women in the ayahuasca community occurs across and within cultures, between Indigenous healers and participants, between Western healers and participants, as well as cross-culturally. However, research and experience indicates that the potential for abuse is further exacerbated by cultural differences in the current context of the increasing globalization of ayahuasca whereby many Western people now partake in ayahuasca ceremonies in

South American contexts or whereby South American healers travel to the West. A main aim of our guidelines is to empower women in these culturally unfamiliar contexts where ayahuasca ceremonies often take place.

Mutual cross-cultural misunderstandings and misconceptions between healers and participants create confusion at least and can be brutally manipulated at worst. Many Western people hold highly romanticized views of shamans and ceremonial leaders, imagining them to be like saints or spiritual gurus. Within their Native communities however, *ayahuasqueros,* as discussed by anthropologists, are viewed as normal men who have varying degrees of healing talents and who do not necessarily occupy esteemed community positions. Yet many *ayahuasqueros* have learned to take advantage of romanticized notions that non-Indigenous people have of them as healers and might use their role to manipulate others for their own personal sexual interests. This often occurs in the context of individual healings called *sopladas* or *limpiezas* where women who are naive about what constitutes usual levels of touching or nudity and are especially vulnerable to abuse. It appears to be common for women to be invited by ayahuasca healers for "special" healing experiences, and then be manipulated or forced in to sexual acts. The guidelines explain that nudity is not typical and shamans do not require their patients to remove undergarments for the purposes of healing. Yet, in a new environment without understanding some basic ground rules, one can be unsure of what is considered necessary or not and find that a boundary might soon slip out of their control. Our hope is that with knowledge of the guidelines beforehand, one can be aware of common manipulative techniques that sexual abuse perpetrators might employ.

A complex and important issue that is raised by the Chacruna guidelines in addressing to sexual misconduct in ayahuasca circles is the issue of mutual consent. Research and experience suggests that many incidents of abuse occur in contexts that can be spoken of in precarious "consensual" terms. Consent lets someone know that sex is wanted but this needs to happen in a mutually intelligible language where "consent" means the same things to the individuals involved. While at the moment of the alleged "consent" all things might seem equal, they often are not. As in any healer-patient dynamic, the healer is in a position of power and responsibility, which creates an unbalance between both parties. Many healers have manipulated vulnerable women in to having sex with them through taking advantage of these uneven power dynamics. Furthermore, often individuals might have no way of knowing that they are being manipulated or influenced by other factors outside of the context of what is meant to be or look like "consent." The presence of ayahuasca in these encounters also raises the question of whether a person can truly consent to sexual relations if under the influence of a psychedelic substance. According to shamanic practice, it is possible for an *ayahuasquero* to influence a woman through shamanic techniques in to feeling sexually attracted to him. Other psychoactive substances have also been used in the wider ceremonial context to decapacitate women in order to confuse and sexually abuse them. It is also common for healers to suggest that having sex with them is a form of healing or a way to gain spiritual power, and deceive women by stating that these relations are morally acceptable to their wives or partners. They might also be given a special position in the ceremonial space to make them feel special or gifted, encouraging them to continue to engage in sexual relations with a ceremonial leader.

Women are often confused and ashamed following these incidents of abuse and feel unable to speak up. The accountability lies with the shaman, who is responsible for resisting the context where this might happen in the first place. On this basis, the guidelines raise the awareness of context for potential seduction that one might wish to consider beforehand. Research also shows that some women stand by their decisions of mutual consensual sex with shamans or their assistants and have no regrets. Some individuals are attracted to the possibility of having sex with a shaman or ceremonial leader and may pursue sexual relations with them. Of course, it is also possible for loving and sexual relationships to be established between ceremonial facilitators and participants in ayahuasca circles. However, as between doctors and patients, it is widely agreed that this is a transgression within the healing context.

The importance of integration is also emphasised in ayahuasca circles, allowing time for the effects of the medicine to wear off and its ensuing sense of empowerment and waiting to "come back down to earth" so that a woman can apply her best judgement as to where she wishes to place her newly found insights. It is the healer or facilitator's responsibility to resist entering into relationships with ceremonial participants until some time after interactions within the healing space. There is no common rule of how long after one should wait though. Indeed, this topic generates heated arguments in the ayahuasca community. Chacruna's purpose with the guidelines is to raise awareness about the complexities of "consensual" sex with an ayahuasca healer so that women can be informed and thus empowered by knowledge. Further discussion is needed across the ayahuasca community and wider psychedelic circles to better establish where the boundaries lie between consensual and non-consensual sexual relations—a conversation that should be ongoing.

Finally, it is important to emphasize that there are many male healers and ceremonial facilitators working with ayahuasca with great integrity who are outraged by sexual abuse in ayahuasca settings. We hope that as well as helping to safeguard women, the guidelines will help to inspire constructive dialogue around sexual misconduct and its elimination. We do not intend to alienate men from this conversation, indeed they form part of our committee. In fact, we believe it is crucial to our communal efforts toward healing that this conversation extends across gender as well as cultural boundaries. Sexual abuse of course affects people well beyond ayahuasca healing contexts. It is a global epidemic in contemporary society. As a community interested in and practicing healing, we are well-positioned to address this grave problem within and perhaps even beyond our community.

SAFETY GUIDELINES

1. *Consider Drinking with Friends.* Partaking in ayahuasca in ceremonies or any healing practice alone with the healer has been a common context in which sexual abuse has occurred. We advise that you consider being accompanied by a trusted companion.
2. *Consider Drinking with Experienced Women or Couples.* As an extra precaution, one may wish to ensure there are female healers or facilitators working in their chosen ceremonial setting. Many reputable places now ensure that experienced women are present to assist and safeguard female participants.
3. *Check Out the Location and Healer.* Check the reputation of any center, shaman, or religious leader you plan to participate in a ceremony with

through review sites, past participants, and other experienced people in the area. It is highly advised to consult women.

4. *It Is Not Necessary for Healers to Touch Intimate Parts of Your Body or Any Area to Which You Do Not Consent.* Some healings are individually focused on the participant's body, such as sopladas (when the shaman blows tobacco smoke over your body; typically head, chest, spine, hands, and feet) and limpiezas or baños de plantas (plant baths, whereby saturated plants are poured over you) but they do NOT entail touching your private parts. If a shaman, religious leader or facilitator does touch you in a way that makes you feel uncomfortable during a "healing," it is your right to assert that you are not okay with this. You can raise the issue on the spot, with trusted facilitators, organizers of the ceremony, or others outside the ceremonial setting.
5. *Curaciones, Sopladas and Limpiezas Do Not Require You to Remove Your Clothes.* It is certainly not necessary for you to be naked. It's true that in certain Colombian yagé (ayahuasca) traditions, it is usual for participants to be asked to remove their shirt for a limpieza, but it is normal for bras or camisoles to be kept on. This is also true for plant baths, for which you can wear swimwear, underwear, or whatever you feel comfortable with. A healer may offer to do a "special" or individual healing outside of the ceremony that can be beneficial, but know that you are free to interrupt or decline any treatment. You may wish to ask another participant or a trusted companion to be with you during any such treatment. You have the right to be assertive about your personal needs to feel comfortable, regardless of any resistance from the healer.
6. *Look Out for Warning Signs That a Healer's Intentions with You Might Be Sexual.* For example, if the healer focuses on your looks, is overly "touchy," he tells you his wife doesn't mind him having sex with other women, encourages pacts of silence and secrecy between you, says he wants to teach you love magic, states that ayahuasca can enhance sexual activity, or declares that you are special and chosen and offers you ceremonial and religious status, beware that these kinds of comments and actions have shown that a healer is likely trying to seduce you.
7. *Sexual Intercourse between Healer and Patient during Ceremonies or Directly after the Ceremonies Is Not Acceptable in Ayahuasca Traditions.*

If a ceremonial leader wants to have sex with you during or soon after the ceremony, he is committing a transgression. This is considered inappropriate and spiritually dangerous in all traditions.

8. *Sexual Intercourse with a Healer Does Not Give You Special Power and Energy.* This is an argument commonly made by men who want to have sex with their ceremonial participants. While no participant can be told what they can and cannot do with their bodily autonomy, sleeping with a shaman will not make you a shaman, heal you from your past traumas, nor give you any special powers or abilities.
9. *Consider Cultural Differences and Local Behavioral Norms when Interacting with Native Healers.* There are some rather benign interactions in Western culture that carry different meanings elsewhere and can potentially be culturally inappropriate and misunderstood. Overt or internalized misogynistic tendencies that view women as being passive—meaning that men merely need to be verbally or physically suggestive with women for sex to take place—are a widespread problem in South America and elsewhere. It may be helpful to consider cultural differences when interacting with healers and their community, as certain behaviors, such as being alone with men, being complimentary, prolonged eye contact, or "free-spirited" behavior, like bathing naked in public spaces, can be misconstrued as gestures of sexual interest. We are not stating that misinterpretation of cross-cultural codes is justified, only that individuals can benefit from being aware of such potential misinterpretations.
10. *Consider Cultural Differences and Local Clothing Customs.* Non-local women are often viewed as being desirable, exotic, and sexually promiscuous across cultures. Without condoning these misconceptions and their underlying assumptions, it may be helpful to consider local clothing customs when attending ceremonies and traveling around in foreign countries. Indeed, the request to not wear revealing clothing is common for many spiritual, meditation, and other healing retreats.
11. *Protect Your Personal Space.* Healers with integrity will respect your right to protect your physical and spiritual space before, during, and after ceremony. You should not feel obliged to engage in verbal or physical

communication with healers, facilitators, or anyone else during or following ceremony.

12. *Be Wary of Healers Who Offer Psychoactive Substances Other Than Those Used During Ceremonies.* The use of additional psychoactive substances within and outside of ceremonies, besides medicinal plants used in the ayahuasca brew and for shamanic dietas, is sometimes associated with scenarios of abuse. These substances may be presented as "medicines" or therapeutic treatments involving the healing of energy imbalances, or "sexual chakra releases."
13. *He's a Shaman, Not a Saint!* Remember, shamans and other ceremonial or religious leaders are men (and women) with human flaws, sexual urges, and the potential to abuse their power and cause harm, like anyone else. They do not necessarily live according to the moral standards one might expect of a spiritual leader. Imagining certain individuals to have superhuman qualities is likely an erroneous and dangerous misconception.
14. *If a Violation Occurs, Get Support.* Don't suffer in silence. It's not your fault if you experience abuse. Ideally, speak out on the spot or let someone in a leadership position within the ceremony circle itself know. However, you may not feel safe to do this or, you may not fully realize abuse has occurred until after the fact. It is very common for women to experience a "freeze" response during a violation or uncomfortable situation. You have the right to report this abuse afterward, even if you were unable to address it at the time. Seek outside support through trusted contacts, and, if necessary, legal advice. Different countries have different legislations; try to get informed about your rights and where an incident can be reported. You may wish to consult the Chacruna Institute's Legal Resource Companion to the Guidelines for the Awareness of Sexual Abuse.[2]
15. *Beware of What Might Appear to Be Consensual Sex.* Consent should happen in a mutually intelligible language where consent means the same things to all individuals involved. If you are considering having a sexual encounter with a shaman or facilitator, bear in mind that this person is in a position of power in that context, and sexual activity may involve an abuse of power. It is also possible, according to some

shamanic practices, for ceremonial leaders to intentionally influence participants into feeling attraction towards them, through love magic and other techniques. Allow time for integration and for the effects of ayahuasca and its often-ensuing sense of empowerment or euphoria to wear off so that you can apply clear judgement.

16. *Beware of Getting Romantically Involved.* Feeling attraction toward the ayahuasquero or a fellow participant can happen. As part of their ceremonial experience, some women have dreams and visions about the shaman or other fellow participants, and can get sexually aroused before, during, and after ceremonies. If such feelings arise, one should not be ashamed of them, but be aware that they may be temporary, and may also be induced through shamanic techniques. Considering the above information and perspectives, pursuing these feelings in concrete terms or not is at your own discretion.
17. *If You Are Aware of or Witness Sexual Abuse, Speak Up!* We are all responsible for combating sexual abuse in our communities. Collaborative efforts are essential to denouncing perpetrators and eradicating sexual abuse in ayahuasca circles.

SECTION SIX

Sustainability, Policy, and Reciprocity

Beyond Prohibition of Plant Medicines

CHARLOTTE WALSH, MPHIL

IF THE PROHIBITION OF PLANT MEDICINES WERE TO END, WHAT DO WE want to come in its wake?

First, I think it's important to be clear about what we want to avoid and why. It's crucial to remember that what change looks like will be strongly influenced by how the case for such change is made. Given this, it's somewhat worrying to me that drug policy reformists often call for change by implicitly buying into the lie at the heart of prohibition. Namely, that the state has the right to control what we ingest, with their rhetoric that, "Yes, drugs are potentially harmful, but prohibition is only making matters worse, so we need to replace it with a system of stringent regulation."

Of course, if we break these hypothetical regulations in the future, we'll still be breaking the law, so we'll have ended up with little more than a watered-down version of the current approach. It's important to be careful what you wish for. We don't want to wrest control from the hands of the powerful, simply to hand it straight back. We need to avoid jumping straight from the frying pan into the fire, transitioning from criminalization to a system of strict governmental legal regulation. Accordingly, we need to resist any framing of the narrative that tacitly suggests that prohibition is an acceptable model and any retraction of it is some kind of favor to us, rather, let's have as our foundational principle an unapologetic assertion that the drug laws are draconian, unsuitable for any civilized liberal democracy, that they need razing to the ground, with a new system built from scratch, rooted in the fundamental recognition of our right to cognitive liberty, to alter our own consciousness.

If we don't stake this claim, we need to recognize that, even if prohibition were to end, all the underground plant medicine healers will most likely not simply be able to emerge blinking into the light of legality, continuing with their work as before, minus the looming threat of criminalization. A lot of people who currently perform incredible work may be frozen out by a system of strict regulation if its requirements are prohibitively expensive or exceptionally tricky to comply with. Indeed, we're seeing this dynamic in action already in the US, with corporate players from Wall Street making a killing out of the legalization of cannabis in some states, as they have the means to abide by demanding governmental regulations, while those from marginalized communities who are still being imprisoned for selling the exact same plant, don't. Not only is this a potent example of social injustice, but it leaves us in an unpalatable situation where people who see cannabis as a commodity are reaping the benefits from legislative change. In contrast, those with a longstanding relationship with what they often consider to be a sacred herb continue to be persecuted. We don't want to repeat this pattern.

Or, how about the fact that, under a system of strict regulation, some sort of license will doubtlessly be needed in order to distribute plant medicines? This raises interesting questions, such as: Who will be in charge of the licensing system? What will it entail? Traditionally, in Indigenous cultures, training in plant medicine work is bound up with what's often conceived of as a divine calling, with strong lineages and lengthy apprenticeships, all of which is a far cry from completing a certifiable course in shamanism. Further, many of those who hold ceremony in the UK are visiting, often Indigenous medicine workers, and how they would fit into any regulatory system is something that will need to be resolved. Suffice it to say that any model that excludes the originators of this work due to a different Western conception regarding what constitutes legitimacy would be profoundly problematic. Issues of power and of authority are palpable here.

Another very real danger is that the prohibition of plant medicines will end, only for them to then be medicalized, falling under the strictures of that system. The plant medicines make for an uneasy fit in the medical model, and we need to think very carefully about whether or not we want them to. Lest we forget, the reason so many of us are turning to alternative shamanic modalities to help with issues such as depression, addiction, and so forth is because Western medicine—for all its undeniably incredible achievements—isn't very good at dealing with malaises of the mind.

The plant medicines are not medicines in the sense in which that term is typically understood in the West. They're perhaps best understood as healing for the soul and thus don't fit easily in to a system where the very idea of the soul has been largely dispensed with. Yet, perhaps this dispensation with the soul is the reason behind Western medicine's failure to deal effectively with mental health issues, most of which I believe are ultimately rooted in spiritual disconnection and thus requires this spiritual dimension to be addressed in order to be resolved. To put it another way, you can't expect to successfully medicalize your way out of an existential problem. The plant medicines can open our hearts back up, hearts that we've shut down and armored off in response to trauma (in response to life!), and it's this re-opening to love that leads to true healing, not just the suppression of symptoms, such as is effectuated by antidepressants.

Of course, we're well along the path towards psychedelics being subsumed into Western medicine, and I'm not saying there's no place for the medicalization of certain psychedelic substances. Indeed, that may be the ideal system for some people to access them, and I'm in awe of and support the incredible work done by MAPS and similar initiatives in the UK moving towards this. However, this is not, and should not be the only way.

While psychedelic psychotherapy is a great improvement on the incumbent model, and seems to work terrifically, with, for instance, MDMA and psilocybin, this clinical approach is not what some people need, and neither is it the appropriate setting for the plant medicines, the effects of which are inseparable from the context in which they're taken. The magic of ceremony is about the medicine, but it's also about so much more: the intention of the individual in drinking, the relationship between them and the healer, the sense of community with others in the circle, the transportive soundscape, and many other factors. Unlike with most Western medicines, this isn't a passive process—sitting back and waiting for the pill to work its magic—but, rather, it takes effort, and we need to do the work. Plant medicines are part of a process that continues as we integrate the lessons we've learned from them. Benefits unfold over time in ways that are often difficult to trace, that are perhaps surprising, as energetic systems are reconfigured as part of a journey that lasts a lifetime.

As another illustration of the incompatibility of the shamanic model with the medical one, Western medicine is rooted in an expectation of predictability of dosage, whereas the plant medicines are inherently unpredictable on this front. What we're witnessing here is a clash of paradigms, with the brewing of ayahuasca, for instance, viewed as a healing art form from the perspective of plant medicine practitioners, with plant spirits very much part of the process—consistency is not what's being aimed for. This illuminates an even more profound conflict between materialistic and spiritual paradigms that cannot be ignored, with those who ingest plant medicines typically perceiving themselves to be in relationship not just with the plants but with plant spirits and beyond. Plant medicine work cannot be forced into the same box as Western biomedicine, as there are entirely different cosmologies at play here. Let's not distort the beauty of ceremony in an attempt to fit it into a model that doesn't work, wrenching these ancestral plants from their necessary context of chaos magic and neutering their inherent wildness in a misguided attempt to render them more palatable. And let's acknowledge that, in resisting this, we're likely to be met with some resistance in return. The medical profession has spent many years establishing a monopoly over the right to provide medicines, wresting this power away from traditional healers, and it's not going to hand that power back over without a struggle.

However, rather than approaching this as a fight, let's acknowledge that there are benefits to both models, that we can learn from one another, creating a truly holistic approach to health, using that term broadly, encompassing the notion of all of us flourishing to our full potential. Compounding these concerns about over-medicalization, recently, we've witnessed what some view as a worrying trend towards the commodification of psychedelic medicine and knowledge by venture-backed companies with somewhat suspicious links to some of the most harm-producing institutions of mainstream society—such companies surely have the plant medicines in their sights.

This leads us to the bigger question of whether we want the plant medicines to be co-opted to take the edge of late-stage capitalism—arguably helping to perpetuate it—or, instead, to fundamentally challenge these systems of injustice that are, in large part, responsible for the traumas we're trying to heal. Fortunately, I don't think the plant medicines are so easy to co-opt or commodify; they're more tricky than that and will doubtless seep beyond the boundaries people try to erect around them. They can help us to reconnect with ourselves, to recognize the illusion of separation from others, to connect with nature, with the sacred, with love, and in doing so, they teach us to find succor from within, rather than endlessly trying to fill that gaping void inside with meaningless objects. As such, they're always, ultimately, going to be a threat, not an ally, to the purveyors of such meaningless objects.

Similarly, in showing us that we're united, they'll always pose a danger to those who seek to divide us. Let's not forget, more broadly, the motive for criminalizing psychedelics on a global level was, at least in part, due to their impact on political consciousness. Now, more than ever, we need to effect social and political change, to come back to love and connection at a time of hatred and division, to recognize our inseparability from nature as environmental cataclysm draws ever nearer, to call on the divine to give us strength to deal with these existential crises.

Let's not underestimate what we can achieve with these plant teachers as our accomplices. Let's not sell them—and ourselves—short. Just as we acknowledge that the value of plant medicines in the domain of healing is that they get to the root of the trauma, it's important to recognize that this trauma is not just individual but, more often, societally induced. As Krishnamurti so eloquently put it, "It's no measure of health to be well adjusted to a profoundly sick society." So, the healing we seek needs to go well beyond the individual, lest we simply mirror the lie of individualism at the heart of the neo-liberal project. Having turned inwards, we need to remember to turn back outwards, to become spiritual activists. And, yes, the plant medicines can catalyze healing, but let's not forget that they're also visionary materials that help us explore different dimensions, at the very least, of our own consciousness and perhaps beyond. From these mystical vantage points, we can hopefully reframe the picture of our reality, bringing back wisdom that will help update outmoded and harmful models of existence.

For all of these reasons, rather than stringent regulation or medicalization, I'm an advocate of decriminalization. To put it bluntly, our right to be left the fuck alone. Decriminalization comes in two flavors: either full, formal decriminalization, which entails the repeal of prohibitive legislation or a watered-down, informal decriminalization, which is normally effected via a change in policing policy, an agreement not to enforce criminal laws that remain in place, most commonly restricted to those that relate to possession in the realm of drug policy. It's full decriminalization that is being advocated for here, the repeal of *all* prohibitive drug laws, which effectively equates with legalization, minus the accompaniment of an enforced regulatory system.

I suspect many people want decriminalization in their heart of hearts but are reticent to say so, believing that we have to give concessions in order to get them. In my ideal world, decriminalization would be accompanied by the rise of plant medicine practitioner groups, drawing up their own guidelines and their own protocols, that could then be voluntarily ascribed to a bottom-up, rather than a top-down model, shaped by those who know what they're talking about. We need guidance from those with wisdom, not law.

My position here is optimistic, maybe even ideological. I make no apologies for that. It's important to argue for what's believed to be right, not simply for what's believed to be possible. We need a clear vision of what self-regulated best practices with plant medicines would look like and to live fearlessly into that. If we shrink our vision by second-guessing what those in power are going to "allow" us to do, we're conceding the game and giving up before we've even begun.

Of course, there's going to be opposition to these ideas; if there wasn't, you can guarantee they weren't ideas worth pursuing. *Of course,* we're going to be met with the rejoinder that what we're proposing is impossible; that's how the status quo responds to threat, by presenting itself as the only viable system, with everything else an impossibility. But what's actually impossible is to carry on as we are doing. And decriminalization isn't just a pipe dream; it's starting to happen. Earlier this year, grassroots campaigning led to the effective decriminalization of magic mushrooms in Denver, Colorado, and, shortly afterwards a smorgasbord of entheogenic plants in Oakland, California.

Furthermore, the world of underground healers is already self-regulating in a disparate way, and more collaborative models of self-regulation exist and are in the process of being developed yet further. Importantly, we don't need to reinvent the wheel here—there are many lessons that can be learned from Indigenous societies regarding the cultural regulation of plant medicines. Let's not do this in a half-assed way, a virtue signaling nod to Indigenous peoples. I'm talking about taking their deep knowledge seriously, taking the idea of communicating with plants and spirit realms literally, rather than metaphorically, recognizing that we've gone astray, that we have so much to learn in order to get back on the path, and that we need to do so swiftly, before we take ourselves down, and the rest of the planet along with us. And, of course, there's an ugly legacy of colonialism and repression of these peoples, including their ancestral medicine practices. Let's not repeat this with spiritual extractivism. We need to think sensitively about the impact that developments in the Global North are having on the Global South, and what we might best do to mitigate any harms that arise. This will involve thinking really carefully about what true reciprocity looks like.

I was at the World Ayahuasca Conference, organized by ICEERS, in Girona this Spring, and the overwhelming message that came from the Indigenous people in attendance was that they needed our support in holding out against the land grabs by corporations in the Amazon that constitute a form of genocide and a form of ecocide. If the plant medicines teach us anything, perhaps it's, through the felt experience of oneness, to expand our circles of concern, leading to a genuine eco-consciousness that respects life in all its forms. Indeed, even operating from a perspective of pure self-interest, this shift needs to happen. The time frame might be different, but we can no more survive without the Amazon Rainforest than the people who live in it. To believe otherwise is recklessly delusional.

In practice, of course, great difficulties will emerge in formulating any such codes of practice. There are markedly different approaches to holding ceremony, for instance, and we'll quickly bump up against exceptionally sticky issues. To illustrate the sensitivity of such matters, consider that, from a Western perspective, it might seem uncontroversial to assume that pregnant women and children will be prohibited from drinking ayahuasca. However, this is certainly not the case in many Indigenous ceremonies or the syncretic churches that have emerged from them, who use them in these cases without evidence of deleterious effects.

What about people with mental health issues? There's a strong argument to be made for robust screening practices, although, who to screen out, and who decides this are very complex questions. Plant medicine ceremonies, almost by definition, tend to involve those with issues in this realm, to greater or lesser degrees, if only by virtue of the fact that they're attended by humans.

So, coming up with models of self-regulation isn't going to be easy, but that's life. It's a never-ending process of negotiating tricky situations. Suffice it to say that any codes devised would need to have a fluidity, a non-specificity to them that allowed for contrasting worldviews by being pared down to first principles.

Fundamentally, we need to create support structures whereby people have access to the best possible knowledge about the plant medicines, where there's skillful preparation, ceremony, integration, so on. We need to ensure that we're always acting in ways consistent with our ethics and broader vision for the future while holding others accountable to behave accordingly.

Indeed, a big question for any collective of practitioners is how to deal with the rogues that inevitably emerge in every field and are perhaps especially dangerous in a setting where they're giving powerful psychoactive materials to (often vulnerable) people. Where these individuals breach the law—in terms of sexual assault, for instance—there are criminal prohibitions in place to deal with this. Where the breach is more in the moral realm, such as not abiding by an agreed-upon code of ethics, there'll be the option of the wider group to first engage with the individual in question. Maybe they follow the principles of a progressive model of transformative justice and perhaps ultimately exclude the person from the collective if they remain non-compliant. Under the recommended voluntary system, this wouldn't stop these providers from holding ceremony should participants still choose to drink with them. Still, the absence or presence of support from the collective would help the process of deciding who to drink with to be a more informed decision with those involved more cognizant of the risks.

Of course, individuals may choose to ingest plant medicines alone or informally with friends, which may (perhaps) be inadvisable, but should certainly not be prosecutable. To put it another way, attempts can be made to provide an appropriate container in which to take the plant medicines; but, should either practitioners or participants choose to step outside that container, that shouldn't invite prosecution, unless, of course, it involves otherwise unlawful acts. *Risk is inherent in life, and there has to be a point at which personal responsibility comes into play in order to ensure freedom.*

It's worth emphasizing that models of self-regulation merit developing both in anticipation of change in the future legal status of plant medicines and regardless of whether or not such change ever occurs. This is it—right here, right now. Not in some imagined future. Indeed, it's always worth reminding ourselves that it's by no means necessary to wait for permission from above to effect positive change from below. Never underestimate the power of underground movements—they change the world. The existence of the plant medicine underground, in all its glory, is testament to this fact: a vast network of—on the whole—advanced souls, doing miraculous work, helping to guide people into the light. I want to end by offering heartfelt thanks to all of those who, under the current prohibitive system, risk their freedom so that we might be free.

A Word in Edgewise about the Sustainability of Peyote

ANYA ERMAKOVA, PHD AND MARTIN TERRY, DVM, PHD

"Do you bite your thumb at me, sir?"
"I do not bite my thumb at thee, sir. But I do bite my thumb."
—Shakespeare, *Romeo & Juliet*, Act I, Scene 1

"SUSTAINABLE USE" MEANS THE USE OF THE COMPONENTS OF BIOLOGICAL diversity in a way and at a rate that does not lead to the long-term decline of biological diversity, thereby maintaining its potential to meet the needs and aspirations of present and future generations.[1]

In the natural sciences, for a long time, discussions about sustainability focused exclusively on the biological aspects. Harvesting of wild plants is sustainable if it is "conducted at a scale and rate and in a manner that maintains populations and species over the long term,"[2] according to The International Standard for Sustainable Wild Collection of Medicinal and Aromatic Plants. A very simplified way of looking at biological sustainability is this: All biological resources are renewable. If the rate of harvest is less than the rate of renewal, then the species or population should persist in the long term.

Recently, this definition has broadened to include the impacts of harvesting on other components of an ecosystem. Harvesting is sustainable when it allows for the long-term persistence of harvested populations and does not negatively affect other species or ecosystem functions.[3] This is a very important condition, as removing one link from the complex web of life can have consequences for the entire ecological community. For example, the flowers of *Banisteriopsis caapi* secrete oils that are collected by specific Hymenoptera species, which are also pollinators. In turn, the ants also tend to and protect the larvae of several lycaenid butterfly species.

But, of course, real life is more complex than this. Sustainability of the harvesting of any natural resource is not a "yes or no" issue, even in the presence of solid ecological knowledge—if that is ever obtainable.

All natural resources used by humans are embedded in complex, social-ecological systems (SESs). Biology is just one of the three components: the ecological, the social, and the economic.

All three need to converge for effective long-term conservation. "Social sustainability" implies cultural appropriateness, social support, and institutions that can function long term. "Economic sustainability" indicates that one activity outcompetes an unsustainable alternative in profit generation.

There are many challenges with estimating sustainability. Natural and social sciences have different approaches and use different concepts and terminology. Moreover, ecological, social, and economic systems are dynamic; so, sustainability criteria for any system are a moving target. How do people begin to understand this multidisciplinary beast? They create models in order to make sense of—and not miss out on—the many intricacies of these interconnected systems.

A common conceptual framework can help to organize findings and provide a map to the complex world of social-ecological systems.[4] One of the most comprehensive conceptual frameworks in the field of sustainability was developed by Elinor Ostrom, the first of only two women to win the Nobel Prize in Economics.

We'd like to present a case study of peyote harvesting in the wild and apply Ostrom's framework in order to assess the sustainability of the current system and to see what the barriers are to achieving maximum sustainability.

CASE STUDY: PEYOTE HARVESTING IN THE USA

Lophophora williamsii (Lem. ex Salm-Dyck) J.M. Coult. (Cactaceae), commonly known as "peyote," is a small psychoactive cactus growing in Texas, the USA, and Mexico. It is a "cultural keystone species": "a species whose existence and symbolic value are essential to the stability of a cultural group over time."[5] In the USA, it is estimated that about 250–500,000 members of the Native American Church (NAC) use peyote as a sacrament and medicine, although it is hard to estimate the numbers with any certainty. Native

Americans fought for many decades to obtain the rights to use peyote but only achieved this as recently as 1994. Peyote is also listed as a Schedule 1 drug in the USA under the Controlled Substances Act of 1970.

The latest assessment from the International Union for the Conservation of Nature (IUCN, 2009) lists this cactus as "vulnerable,"[6] due to the fact that an ongoing population decline of ca. 30% per annum has been projected. In 1995, the NAC declared a "peyote crisis," due to increasing difficulty in obtaining peyote, as well as the diminishing size of peyote buttons. In the last 20 years, the situation has only worsened. In the USA, the subject of the peyote population decline has been raised multiple times since the 1980s. Harvesting is only one of the threats facing peyote. There is also large-scale land use change and the unknown effects of climate change.

This exercise is going to appear reductive and will look at peyote from a different perspective than what you're used to. You might think of peyote as a drug, as a medicine, or as a sacrament, so counting cacti and calling them "resource units" is weird. Our lens is that of conservation biologists, and we view peyote as a vulnerable cactus in peril. By temporarily looking at it as a natural resource, we can apply what we learned from other challenging conservation problems. What we'd like to see is a move beyond thinking about peyote in solely one category or another, whether it is a natural resource, a trade commodity, a sacrament, a medicine, a Schedule 1 drug, etc. It is all of these and more. We need a more integrated approach to achieve sustainability.

The goal is twofold:

- **Ecological Performance Measures:**
  Peyote biological sustainability and the overall ecosystem health/functioning

- **Social Performance Measures:**
  Guaranteed access to peyote for religious, medicinal, and cultural purposes for the members of the Native American Church

We cannot go into the details of every indicator as there are 56 of them, and each one could be a topic of its own article. However, most of the indicators show that we do not have enough data, and the data that there are indicate that harvesting from the wild, given the current methods and regulatory measures, is not sustainable.

What Do We Need in Order to Understand Sustainability?

- The rate of growth and renewal is a vital knowledge requirement, as is a species' response to different harvesting regimens. This is key information for defining an ecologically sustainable yield, e.g., sustainable yield = a harvesting rate < than the renewal rate. Terry et al. made a reasonable start on this, but we need to understand the effects of different harvesting practices and harvesting rates as well.[7]
- Development of rapid, reliable, and credible peyote inventory techniques.[8]
- Clear, evidence-based guidelines on harvesting quotas and techniques. Sustainable harvest systems must be flexible and adaptive; i.e., harvest rules or quotas should not be fixed but rather consistently monitored, evaluated, and adapted to the current conditions. We've made initial recommendations, but these need to be applied and tested in practice![9]
- Understanding of the dynamics of the ecosystem and its interaction with the plant.
- Knowledge of species abundance in the environment, the factors that drive their abundance and productivity, and their responses to harvesting and other external pressures.
- Understanding seasonal changes, environmental variables, and response to stochastic events, such as drought, floods, freezes, pests, etc.
- Establishing refugia, or some safety nets for peyote. These can be size refugia (do not harvest large seed producers or small plants); or spatial refugia (presence of populations in the areas inaccessible to harvesters, such as protected areas).
- Impact of peyote harvesting on other elements of the ecosystem. How does the removal of peyote cacti affect the remaining members of the biological community? We simply don't know!
- Understanding interactive effects of harvesting and other drivers, including climate and land use changes.

But, more urgently than all of the above, we need consultation with and respect for the beliefs, wishes, and opinions of the key actors: Native Americans and landowners.

The landowners should also be an integral part of any initiatives, as peyote grows on their property, and they are—or could be—the main custodians and stewards of the land. As part of economic sustainability, there should be incentives for ranchers to protect peyote on their land, something that can convince them that conserving Tamaulipan thornscrub is more important than root-ploughing it for agriculture of "improved grazing," or leasing it to build oil, gas, and wind-farm infrastructures.

Native American Church members are the main consumers of peyote and absolutely have to be involved and listened to when developing any conservation initiatives. We need to understand the attitudes and develop alternatives to harvesting peyote in the wild, but only if these alternatives are culturally acceptable.

With respect to harvesting practices and guidelines, the development of effective and locally appropriate participatory monitoring systems is needed. What is the purpose of the guidelines if no one follows them? The governance systems are the key to ensuring sustainability. But, to be effective, they need to be ecologically informed, adaptive, and flexible (via feedback loops provided by monitoring and evaluation), which is absolutely not the case in the current situation. Experience from many sectors all across the world indicate that community-led initiatives are more sustainable in the long-term than top-down hierarchical governance. Managing complexsocio-ecological systems requires institutional arrangements that are more flexible and responsive to rapidly changing conditions than top-down governance regimes. Under current regulations, re-wilding[1] and habitat restoration—not to mention cultivation—initiatives involving peyote are akin in their regulatory complexity to manufacturing a Schedule 1 drug such as heroin! Moreover, the current regulations in place do not seem to be effective, and poaching in South Texas remains a serious, continuing problem.

CONCLUSION

The sustainability question is an amalgamation of cultural, biological, economic, political, and social elements—all important and often indistinguishable. Solid ecological or environmental knowledge and evidence-based solutions are necessary but not sufficient to establish long-term sustainability for the species. Sustainability necessitates the integration of cultural perspectives and scientific discoveries to tackle challenges, with the additional strand that weaves these together: environmental and social justice—justice, not limited to natural resources, but also including rights to cultural diversity in terms of knowledge and practices.

(*) Rewilding is large-scale conservation aimed at restoring and protecting natural processes and core wilderness areas, providing connectivity between such areas, and protecting or reintroducing apex predators and keystone species.

It's Time for the Psychedelic Renaissance to Join the Harm Reduction Movement

GEOFF BATHJE, PHD, VILMARIE FRAGUADA NARLOCH, PSYD, AND JOSEPH RHEA, JD, PHD

THERE HAS BEEN A BOOM OF INTEREST IN PSYCHEDELICS, THOUGH MANY people, including licensed providers (LPs), are unfamiliar with the applicable laws and gray areas. Supports, safety nets, and drug knowledge still need to be expanded for those who use psychedelics, while the critical analysis of where psychedelics fit into the broader War on Drugs often seems lacking or underappreciated. As clinical research on psychedelics as medicines has progressed and awareness of their therapeutic potential has risen, LPs must decide how best to support their clients who use these substances. This includes discussion of the role of harm reduction and "integration" in relation to psychedelics, along with possible legal and ethical considerations for LPs in providing these services. At the same time, LPs, researchers, and advocates must grapple with how to address psychedelics in a way that does not reinforce the harms and injustices of the War on Drugs. We believe the harm reduction movement has much to offer to those who are interested in the current "Psychedelic Renaissance."

THE HARM REDUCTION MOVEMENT

Harm Reduction is a well-established approach to reducing the negative consequences associated with drug use. It is also a social justice movement, rooted in respecting the rights and humanity of people who use drugs. Thus, harm reduction includes not just reducing risks of drug use, but engaging in policy work, community organizing, and activism to reduce stigma and harmful drug policies. The harm reduction movement dates back to the 1920s in the UK with the Rolleston Report Committee recognizing that drug use need not be criminalized,[1] and to 1940 in Mexico, when legislation allowed doctors to prescribe drugs to people with addictions and removed criminal penalties for possession.[2] In the US, harm reduction practices surged in response to the AIDS epidemic, leading to the development of grassroots movements such as the Gay Men's Health Crisis, Act Up, and Stop AIDS.

Essentially, harm reduction emerged from these communities not being well served by traditional health care, who took matters into their own hands. This tradition continued as syringe exchange and naloxone dispensing programs were propagated by people who use drugs and their allies, often initially operating illegally or in the gray areas of the law to save lives and protect health. This level of policy work to identify and navigate the gray areas in the law is not well suited for the average individual licensed practitioner (LP), but instead tends to emerge out of collective efforts to find creative ways to provide resources and services while averting criminalization.

The work of the late Alan Marlatt and colleagues from the University of Washington, Patt Denning and Jeannie Little at the Center for Harm Reduction Therapy in San Francisco, and Andrew Tatarsky and colleagues at the Center for Optimal Living in New York City, and others have further extended harm reduction into therapy settings and provided a theoretical basis for the practice of harm reduction psychotherapy.[3] This approach to therapy maintains the original spirit of harm reduction in recognizing that even small behavioral changes like not sharing syringes can reduce or even eliminate certain risks such as HIV transmission. As such, this approach involves supporting clients in making "any positive change," as they define it for themselves.

PSYCHEDELIC HARM REDUCTION

Historically, the psychedelic and harm reduction movements have not overlapped significantly. However, we believe the sociopolitical lens and praxis provided by the harm reduction movement is very relevant to the current "Psychedelic Renaissance." For example, the most visible and powerful figures in psychedelics are disproportionately highly educated or wealthy White men. What are the consequences of such a homogenous group making most of the decisions and holding most of the power? Should it be surprising that we have seen increasingly frequent criticism of inequality, power, lack of diversity, and lack of systemic critique in psychedelics?[4] From a harm reduction perspective, those most at risk and most in need should hold power and receive the most attention. People who use drugs often form their own organizations or are sought out for prominent roles in harm reduction, such as community organizers, program developers, outreach workers, or board members.

Beyond these systemic issues, it is often said that negative attention resulting from risky psychedelic-related incidents in the 1960s was used to discredit research and implement prohibitionist policies. Proponents of the War on Drugs will always seek to stigmatize use and publicize the most negative experiences to denounce and impede progress toward liberalizing drug policy. Harm reduction provides an approach to drug use and drug policy that can help prevent similar outcomes with psychedelics in the current era.

Despite the hesitation of some within the psychedelic community to acknowledge the risks inherent in the use of psychedelics or to draw parallels between medicinal and personal use, there are perils to psychedelic exceptionalism (viewing psychedelics as inherently superior to other substances, or deserving of special legal protection). For example, exceptionalism can lead to the stigmatizing of other drugs or people who use them by negative comparison, advocating for only psychedelics while neglecting those most oppressed by the War on Drugs, discounting that almost any drug has medicinal uses and that much drug use is for symptom relief, or neglecting that various cultures have considered a range of drugs to be sacred (e.g., tobacco, wine, opium, coca, cannabis, etc.). Still, harm reduction efforts for psychedelics have proliferated in recent years. Organizations such as the Zendo Project, DanceSafe, Bunk Police, Energy Control, Students for Sensible Drug Policy, and others provide much needed peer support, drug education, and other harm reduction services (drug checking, safer use materials, etc.). However, many of these efforts have focused more on events where psychedelics are consumed than on individuals using in more private settings, the broader social justice goals of the harm reduction movement, or on LPs providing harm reduction in individual or group therapy settings.

LICENSED PRACTITIONERS IN HARM REDUCTION

There are several instances where harm reduction may be relevant to LPs working with clients. While being overly conservative would inhibit much harm reduction work, all professional ethical codes emphasize the necessity of competence in practicing any approach.[5] So LPs should seek appropriate training and familiarize themselves with the aforementioned references (as well as general discussions on providing psychedelic-assisted therapy prior to it being legally available) if they intend to provide harm reduction services to people who use drugs.[6]

An important role for LPs providing harm reduction therapy involves helping individuals identify and prevent potential harms associated with their use of drugs with a focus on "drug, set, and setting." For our purposes, we can broadly define "set" as things about the individual (such as mood, personality, personal identity, health status, etc.) and "setting" as things in the person's environment (the social and cultural context). People, of course, also need accurate information about the drugs they use, including their effects and impact of the dose size and how to reduce the unique risks associated with each drug.

Harm reduction can include a range of other activities, such as explaining about testing drugs for impurities (though cautioning clients is warranted in states where drug testing kits are considered to be paraphernalia), explaining laws related to drug use, and paraphernalia, or encouraging safer practices for use. This work requires the therapist to have a broad knowledge of various drugs and their use. Multiple resources can be found for each drug, such as the Harm Reduction Coalition's "H is for Heroin" and related guides, or Students for Sensible Drug Policy's "Just Say Know"[7] program, which covers DMT, ketamine, LSD, psilocybin, and cannabis, among other drugs. While we cannot fully vouch for the accuracy of every guide, these guides are generally valuable for LPs and people who use drugs.

As LPs explore psychedelic harm reduction with clients, they should be aware of the limits of First Amendment protection. In Conant v. Walters, the Ninth Circuit Court of Appeals examined the situation of California doctors recommending cannabis to their patients.[8] The Court found that those recommendations were protected from federal action (DEA licensing revocation and federal investigation) because California doctors were not prescribing cannabis or in any other way directly involved in providing cannabis. There is thus a line between free speech and discussion (which courts are very reluctant to restrict) and actions, such as providing contacts for the procurement of a controlled substance or holding a space open for the use of a controlled substance. These latter activities expose LPs to criminal prosecution for conspiracy. At the state level, we are unaware of any licensing actions based on an LP engaging in a discussion with a client about the client's use of psychedelics.

When clients express an intention to consume a psychedelic or other drug, LPs are not required to dissuade them from doing so. In fact, within the harm reduction approach, such attempts to control behavior are seen as unhelpful judgments that can produce stigma and shame, make it difficult for clients to be honest, create resistance that might increase risky behavior, weaken the therapeutic relationship, and increase the odds of clients terminating therapy. When a client expresses they are using or intend to use a drug, harm reduction-oriented LPs will tend to:

1. Seek to understand the function of the drug for the client (What does it do for them? What do they get out of it?).
2. Affirm the client's autonomy to make that choice, so as not to promote resistance in the relationship.
3. Practice radical acceptance and refrain from judgment of the behavior.
4. Ask clients if they are aware of the risks of their drug use and how to protect themselves (offering accurate information if the client is receptive).
5. Ask the client whether they dislike or are seeking to change their substance use in any way while respecting their goals.
6. Offer, but not require, other approaches to achieve the same goals the client hopes to achieve by using drugs (e.g., offer trauma therapy to address flashbacks that the client manages with heroin).

WHERE DO PSYCHEDELIC "INTEGRATION" AND "PREPARATION" FIT IN WITH HARM REDUCTION?

Some in the field are reluctant to support harm reduction and integration services, which seems to be related to confusion about the scope and definition of each. Each term has been used to denote a range of LP conduct: from recommending psychedelics to clients (clearly illegal outside of sanctioned uses, such as clinical trials or ketamine clinics) to merely counseling people who have independently chosen to use a psychedelic (clearly legal).

We define "integration" as the processing one provides after a psychedelic experience. This typically involves discussing and internalizing experiences a person believes were beneficial to them (e.g., a sense of reconnecting with oneself or the lifting of one's depression) that were negative (e.g., a "bad trip") or that they have not yet been able to interpret (e.g., what did a particular part of their experience mean?). "Harm reduction," on the other hand, mostly refers to things done prior to a person's ongoing next use of a drug, such as discussing a person's pattern of use and providing information intended to reduce risks (e.g., encouraging them not to use while driving or in other high-risk situations, discussing the impact of set and setting, etc.). Warning someone about the dangers of a behavior is unlikely to ever be considered legally or ethically problematic.

Harm reduction and integration overlap with regard to reducing the harm or ongoing risks of difficult or negative experiences with psychedelics (e.g., helping someone through heightened anxiety or suicidal ideation stemming from a difficult psychedelic experience). Harm reduction and integration can be further differentiated from "preparation" for therapeutic experiences. In preparation, LPs who do not provide the drug or a physical location for its use represent that their professional services can (in advance of use) make an illegal drug experience therapeutic. The risk here is that if an LP becomes a material accessory to the drug experience during preparation, a client could potentially claim the only reason they followed through with the use of an illegal drug was because of the involvement of the LP. In that case, a licensing board may determine the LP made themselves into an integral part of the illegal experience under the guise of providing professional services (and violating duty of care). In summary, LPs face greater risk when doing preparation work versus post-experience integration, particularly when the provider is recommending (or is perceived as recommending) the psychedelic experience as a therapeutic activity as opposed to focusing on preventing harm.

Some have warned of the legal risks of integration work based on it being "new" or unestablished. But, despite the diversity of techniques that are described as integration, from being in nature to journaling, art, or therapy, the concept is not new in psychology. The earliest reference to integration in psychology may be Pierre Janet's definition in 1889 of traumatization as a "failure of integrative capacity,"[9] with the major goal of therapy being the restoration of the ability to integrate experiences. Further, the integration of insights derived from therapy has always been implicit in psychoanalysis, if only through sheer repetition and time spent in therapy. As briefer models of therapy were developed, most theorists took for granted that clients would retain things they worked through in therapy. As a result, relapse rates for most conditions are exceedingly high. Of course, even in brief therapy, therapists often recognize that much of the work of therapy occurs between sessions and supplement client processing and integration by recommending self-help literature and "homework" to be done between appointments. Further, some theories of psychotherapy have explicitly put a major focus on integration (see Sensorimotor Therapy, also known as Sensorimotor Processing and Integration). It is false to claim that integration is a new and untested approach, as the concept of integration is ubiquitous throughout different schools of psychotherapy, even if described in different terms.

CALLING FOR MORE HARM REDUCTION

The American Psychological Association's (APA) ethical guidelines and the guidelines of many other LPs support practice based on a range of evidence, following one's conscience, and prioritizing the well-being of clients. According to the APA, "Psychologists may consider other materials and guidelines that have been adopted or endorsed by scientific and professional psychological organizations and the dictates of their own conscience, as well as consult with others within the field."[10] Of particular note is that harm reductionists are often guided by their conscience in feeling an obligation not to deny people therapy just because they have consumed drugs that are currently illegal, especially when they believe these laws exist to reinforce classism and racism. Certainly, there is no legal or ethical requirement to turn such clients away or deny them the opportunity to process their experiences with drug use, regardless of whether they see their drug use as beneficial, problematic, or neutral.

In conclusion, it is not only appropriate but essential for LPs to learn about and engage in harm reduction related to psychedelic drugs if we wish to protect our clients and remain part of the broader justice movement against the War on Drugs. We would be naive to think everyone who is struggling with depression, trauma, addiction, and other intractable issues will wait years for medical approval of psychedelics before trying them on their own. The positive publicity around the therapeutic potential of psychedelics is already causing many people to seek out experiences they hope will be healing. Of course, harm reduction is not only for LPs. People who use drugs and their allies have always been at the forefront of harm reduction, and it is important that they continue to be so. We call on LPs who are interested in psychedelics as medicines to join and learn from the broader harm reduction movement.

What Do Psychedelic Medicine Companies Owe to the Community?

MATTHEW BAGGOTT, PHD

VENTURE CAPITAL IS BETTING BIG ON THE POTENTIAL OF PSYCHEDELICS to transform into lucrative medicines. More than half a billion dollars have been poured into for-profit psychedelic ventures. Yet, natural psychedelics come from the humblest of origins. María Sabina, the *curandera* who revealed the psychedelic mushroom to the Western world, ultimately died in poverty. As psychedelics are made into corporate medicine, are we repeating a shameful history? What lessons can her history teach us?

The discovery of psychedelic mushrooms by the West can serve as a parable to help us understand what psychedelic medicine companies owe to the community. I will use this history to illustrate some ethical and pragmatic reasons why psychedelic users should be recognized as stakeholders by companies developing psychedelic medicines. One implication of this is that formal structures and processes are needed to include these stakeholder interests in the work of psychedelic companies so that we can together create truly transformative healthcare.

HOW THE WEST WAS SHROOMED

In the summer of 1955, a vice president from the J. P. Morgan Bank and a New York fashion photographer made history by becoming the first outsiders to participate in a sacred mushroom ceremony. The ceremony was held by the locally-respected curandera María Sabina in Oaxaca, Mexico.

The experience shook the banker to his core. Gordon Wasson saw, more clearly than he ever could with his mortal eyes, the newness of everything, as if the world had just dawned. It overwhelmed and melted him with its beauty. And, all the time that he was seeing these things, the wise woman María Sabina danced and sang, not loudly, but with authority.

Wasson had gone to Mexico seeking a religious experience, and he was not disappointed. For the first time, the word "ecstasy" took on real personal meaning for him. For the first time, ecstasy did not mean someone else's state of mind.

Wasson seems to have been a genuine seeker with good intentions, yet his religious exaltation was achieved through questionable power dynamics and subterfuge. Sabina had agreed to meet with Wasson and conduct a ceremony because she had been asked to by a local authority. She did not believe she had a choice.

Once they met, Wasson attempted to mislead her in order to be allowed to partake in a ceremony. María Sabina primarily held her ceremonies to cure the sick. Wasson, therefore, told her what must have seemed to him a white lie, that he was concerned about the health of his son. No doubt, he was as concerned as any parent would be when traveling without their child. But traveling parents usually aren't concerned enough to see a doctor or priest.

After he returned to New York, Wasson described his adventures in a now-famous article in *Life* magazine. Wasson gave María Sabina a pseudonym to lightly disguise her. However, I know of no evidence he warned her of his intentions to publish, nor does it appear he sought permission to publish photos showing her face.

The 1957 article coined the term "magic mushroom" and helped open the floodgates of the psychedelic revolution. People began coming to Mexico to experience God. To Jonathan Ott, the mushrooms were rescued from oblivion at the moment when their use had almost disappeared.[1]

The effects of the psychedelic revolution on María Sabina were mixed at best. She became well known and was visited by musicians, poets, and celebrities. Yet, her house was burned down, she was raided by *federales*, and she was forced, for a while, to leave her hometown. She saw others start offering psychedelic mushroom experiences for money without following the traditional ways. María Sabina later described the results with heartbreaking matter-of-factness:

"But from the moment the foreigners arrived to search for God, the *niños santos* lost their purity. They lost their force, the foreigners spoiled them... Before Wasson, I felt that the *niños santos* elevated me. I don't feel like that anymore."

María Sabina died in 1985 at the age of 91. She had never owned shoes.

LESSONS FOR TODAY

Today, big money is seeking big returns by developing psychedelics as products. Will prestigious bankers again prosper while curanderas remain barefoot?

It may seem hyperbolic to compare unsanctioned users of psychedelics to María Sabina. Yet, the history of María Sabina is still relevant and illustrates two important truths. First, the communities that originate psychedelic practices need to be considered in ethical calculations because they may experience adverse effects from commercial or research activity. Second, a crucial part of the therapeutic power can be lost in translation when a psychoactive substance is removed from its cultural context.

PRINCIPLES FOR PARTNERING WITH SOURCE COMMUNITIES

In modern terms, Wasson was exploiting indigenous knowledge without getting adequate consent when he publicized María Sabina's ceremony. Even considering the context of his time, we can say he should have been less coercive and more honest in his actions. He seemed to recognize this in 1976 when he wrote that he regretted how his actions affected Sabina. But, even when he wrote those regrets, it was still not too late for him to intervene; yet he did nothing.

What should Wasson have done? The only way to know the right answer would be if he had asked. And this is the core of the problem: Wasson treated María Sabina as a resource instead of as someone who should be allowed to make decisions for herself. Ultimately, he should have treated María Sabina—and the residents of her village—as partners. In this section, I discuss how adequate partnering requires "Free, Prior, and Informed Consent" (often abbreviated FPIC) and how entire communities should often be recognized as stakeholders.

Free, Prior, and Informed Consent has emerged as an idea in work on the rights of Indigenous peoples to self-determination. It means that a non-coercive process should be used in advance to inform a group and allow them to choose and direct their participation in a proposal that uses their resources or otherwise affects them. This overlaps with the practice of informed consent in research ethics. Ethical research requires that participants understand study procedures, risks, benefits, and alternatives before freely agreeing to be in a study.

A key difference between the ideas of FPIC in Indigenous human rights and informed consent in scientific research is the focus on community in the first and individual in the second. However, this difference is shrinking. While clinical research ethics has traditionally focused on individual participants, there is increasing recognition that non-participants can also be harmed by research. A study that identifies a genetic risk of developing Parkinson's disease in participants will also suggest the same risks exist in relatives, who may not wish to have this knowledge revealed to themselves, let alone others. And studies that withdraw people from HIV medicines in order to test a new potential treatment can expose sexual partners of the participants to increased risk of contracting HIV. Scientists designing these studies, therefore, need to consider potential benefits and harms to non-participants.

It is perhaps easy to see how Wasson might have overlooked the possible effects of his actions on María Sabina's community. Understanding a community's stake in some piece of knowledge or other resources can be difficult. Dutfield has pointed out that local use of traditional knowledge is often much broader than what is readily reflected in economic exchange value.[2] We see this with Maria Sabina's example, where there was a simultaneously increased market for mushroom trips along with a harder-to-quantify loss of sacred quality. These different uses that a community has for some knowledge or other resource make the community vulnerable to changes in the resource and thus give the community a stake in what happens to it.

If we agree that the local uses of a resource give the community some stake in the resource, it seems to follow that psychedelic users have some stake in how psychedelics are developed as medicine.

One aspect of this is that psychedelic users have, essentially, served as human guinea pigs, elucidating the risks, safety profile, and efficacy of psychedelics. Groups developing psychedelic medicines have often explicitly argued that recent and traditional human use of psychedelics means the drugs are physiologically safe and more likely to be approved as medicines than most experimental compounds. Much of this psychedelic use has been unsanctioned and illegal, placing the user at risk of loss of liberty or worse. For example, Alexander Shulgin—who invented several of the psychedelics now being studied and was one of the first to recognize the therapeutic potential of MDMA—lost his laboratory license and was harassed by the Drug Enforcement Administration because he was about two decades too early in publishing books about the benefits of psychedelics.

These relatively visible individual users are supported by and support less-visible economies and communities. These economies include those who grow, manufacture, transport, and sell psychedelics. These communities also include those who come together to use psychedelics in religious, mystical, or life-affirming ways and those who facilitate growth or healing by conducting psychedelic therapy, integration sessions, or wellness retreats.

Psychedelics are somewhat unusual among illegal drugs in that the motives of many of those involved in commercial activities are not primarily financial (e.g., Henderson & Glass, 1994).[3] Often, the main motivation of people involved in psychedelic economies is to make the world a better place. Think Sunshine Makers, not drug dealers. Now, this doesn't necessarily mean the people maintaining these economies deserve special consideration over those who are motivated by money. But it should suggest that these psychedelic economies are fragile and have meaningfulness that cannot be readily replaced by a legal, commercial marketplace. It is unclear how psychedelic medicines would affect these economies, and it is important to consider.

As a concrete example of this, we might consider underground psychedelic therapists and guides. These individuals have engaged in illegal practices that, nonetheless, have helped many people. With the approval of psychedelic medicines, the skills and capabilities of these underground therapists will be greatly needed. Yet, there is a real risk that many will be shut out of the medicalized system of legal therapy. The US Food and Drug Administration is currently arguing that psychedelic therapy teams should include one person with a clinical doctorate, something most therapists have not needed. Companies developing psychedelic medicines should resist building new systems that needlessly exclude community expertise. More than that, companies should proactively seek to identify and strengthen existing community systems that support psychedelic-assisted wellness.

Overall, while there may be no magic, one-size-fits-all process for appropriately using knowledge from another community, there are broad principles and increasing recognition that considering impacts on the community is the ethical and right thing to do. Companies developing psychedelic medicines should recognize psychedelic users as stakeholders and create processes for including them. This could involve including these communities in early fundraising rounds. Companies could employ structures, such as cooperatives and public benefit corporations, that open the organizations to more stakeholders than just those who own shares. Deciding on these actions and structures should be done in dialogue with psychedelic communities.

LOST IN TRANSLATION

Treating psychedelic users and communities as important stakeholders is not just an ethical issue, it is also a pragmatic recognition of the roles they play in maintaining health. For María Sabina, the mushrooms were not really medicine in a contemporary Western sense of the word. They did not heal like a medicine, they diagnosed and revealed truths that led to healing. The mushrooms told her the origins of any sickness and what therapeutic intervention was needed. It was then up to the curandera and her patient to do the healing.

This was largely lost on the Westerners. Wasson willfully misinterpreted what he experienced. He was looking for mystical ecstasy and clues to the origins of religion, not whether his child was sick or lost.

Today, we risk making similar mistakes. We don't just need better, more powerful drugs; we need people in the community who can help us lead healthier lives. When research focuses on the ability of psychedelics to trigger mystical experiences, it risks glorifying brief moments of chemical ecstasy while minimizing the need to actually change and improve one's day-to-day life.

This change is hard to do on one's own. We need help integrating powerful experiences. This can take the form of formal integration therapy, or it can be just talking to others who understand what we're experiencing and can point us in helpful directions. These practices can go beyond supporting individual health and can help strengthen community resilience.

The success of psychedelic medicine will ultimately be dependent on community structures that adequately support efforts to change and heal. Ironically, Wasson's efforts disrupted these community structures in Mexico. He made mushrooms more available but healthcare less effective. Let's not repeat that mistake today.

Sacred Reciprocity: Supporting the Roots of the Psychedelic Movement

CELINA DE LEON, MDIV

I IMAGINE THE DAY WHEN FAMED HARVARD ETHNOBOTANIST RICHARD Evans Schultes came upon the Sibundoy Valley and the interaction that he had with Salvador Chindoy when he took a picture of Salvador's nephew holding a *Brugmansia* flower. This became a memorable image in *Plants of the Gods*, a foundational text for many Westerners about botanical psychedelics. Schultes would later become a major contributor to expanding awareness about yagé, or ayahuasca.

Similarly, I imagine that day in Huatla de Jiménez, Mexico, when Gordon Wasson and his wife Valentina Pavlovna, amateur ethnomycologists, met María Sabina for the first time and discussed the healing powers of *los niños santos*. She allowed Wasson to take her picture and asked him not to print it, but he did anyway. *Life* magazine later featured the story, and with that, the magical powers of psychedelic mushrooms became known to the rest of the world.

We may consider these both origin stories of the psychedelic movement as it exists today. Of course, there are many others, but these are definitely among them. Many other scholars, scientists, and seekers followed in the footsteps of Schultes, Wasson, and Pavlovna. The knowledge—which had been safeguarded for generations among the humble confines of Indigenous communities—became the foundation for an entire cultural movement.

The newest rendition of this is more commonly referred to as the "modern psychedelic renaissance." As is common in periods of advancement and breakthroughs throughout history, the glory that prevails in one area often exists alongside gaping inequity and profound loss in other areas.

In this case, the movement is not to blame, per se, nor may we consider it a direct transgression of malintent or conscious colonialization. However, at pivotal moments, such as the one we are now living in, we may embrace a simple invitation to reflect upon the current state of affairs. Indigenous and traditional plant medicine communities are suffering, and some are on the very real verge of complete extinction, due to the multi-generational impact of systematic colonialization and genocide. This is a true and simple fact.

In both Sibundoy, Colombia, and Huatla, Mexico, the Kamentsa and the Mazatec are living examples of communities facing very real challenges. These communities are some of the original stewards of psychedelic medicinal plants and fungi, and yet they have benefited very little from the current expansion of interest in psychedelic medicines throughout North America and Europe. While millions and millions of dollars are poured into research and commercialization, the communities where these sacraments originate are severely threatened in multiple ways. Traditional culture and language are being lost, traditional land is being encroached upon by development interests, abject poverty is rampant, and their governments do not offer sufficient support nor adequate resources.

To get a more specific sense of the scale, Compass, a mental healthcare company interested in psilocybin research for treatment-resistant depression, recently announced that they raised 80 million dollars, thus creating a total of 116.2 million raised in three rounds of funding.[1] There are many companies like Compass that are rapidly entering the psychedelic space. Sadly, all of these endeavors have at least one aspect in common—they all lack any clear channel for reciprocity toward Indigenous and traditional communities where these practices originate.

This story is not new. What can be new is how this story unfolds going forward and how the psychedelic movement at large can consider an approach of engaging in meaningful reciprocity by honoring the roots upon which the movement originates. The reality is that the majority of pioneers in the field of psychedelic research, commercialization, and medicalization benefitted in some way from the knowledge base that came from Indigenous and traditional practices of using plant sacraments.

The very protocols that exist today draw upon many of the perspectives and insights that leaders in the field experienced while engaging with these substances, either directly with Indigenous communities or in so-called "underground" settings. These insights and healing experiences, in addition to traditional perspectives on how to use these substances for healing purposes, were often foundational in directing inspirations toward psychedelic research and funding. There is nothing inherently wrong with this, but if there was ever a time to look at this honestly and value the importance of reciprocity, that time is now. It's possible to write a new narrative of a Western movement that engages in meaningful reciprocity with Indigenous culture, cherishing the roots that allowed it to flourish.

Indigenous and traditional plant sacrament communities that have stewarded this sacred knowledge for so long need our engagement—not with an arrogant savior mentality, typified by an attitude that they require our help in the form of pity or charity. Rather, this is about an act of solidarity and social responsibility. Indeed, we are in a true psychedelic renaissance. The potential for these sacraments and substances to positively impact humanity as a whole and help alleviate human suffering is nothing short of awe-inspiring. At the same time, we may consider shifting the ethical compass in the movement at large to one of more balance. We may consider honoring the very principles that psychedelics so often teach us, namely, the significance of our interconnection and the importance of reciprocity.

Directly supporting initiatives that engage in meaningful reciprocity is a great way for individuals, foundations, and businesses to be part of the change. Supporting projects that are Indigenous-led or facilitated by individuals that have long-standing relationships in Indigenous and traditional plant medicine communities is a great step. We must be honest and sincere about the true need for meaningful impact by way of supporting *long-term* sustainability, *long-term* resiliency, and *long-term* relationship building. Now is the time to bridge good intentions with action. Now is the time to consider that supporting this effort is not about a one-time charitable donation but, rather, about investing in and actively creating a world that we want our children to live in—a world that respects diversity, equity, balance, and reciprocity.

Contributor Biographies

EDITOR BIOS

Beatriz C. Labate, PhD

Beatriz C. Labate is a queer Brazilian anthropologist based in San Francisco. She has a Ph.D. in social anthropology from the State University of Campinas (UNICAMP), Brazil. Her research focuses on the study of plant medicines, drug policy, shamanism, ritual, religion, and social justice. She is the Executive Director of the Chacruna Institute for Psychedelic Plant Medicines. Additionally, she serves as the Public Education and Culture Specialist at the Multidisciplinary Association for Psychedelic Studies (MAPS), and Adjunct Faculty at the East-West Psychology Program at the California Institute of Integral Studies (CIIS). She is author, co-author, and co-editor of twenty-two books, two special-edition journals, and several peer-reviewed articles.

Clancy Cavnar, PsyD

Clancy Cavnar is a clinical psychologist, artist, and researcher based in San Francisco. She has a doctorate in clinical psychology from John F. Kennedy University in Pleasant Hill, CA and is Co-Founder and a member of the Board of Directors of the Chacruna Institute for Psychedelic Plant Medicines. Additionally, she is a research associate of the Interdisciplinary Group for Psychoactive Studies (NEIP). She has a master of fine arts in painting from the San Francisco Art Institute, a master's in counseling from San Francisco State University, and a certificate in Psychedelic-Assisted Therapies from the California Institute of Integral Studies (CIIS). She is author and co-author of articles in several peer-reviewed journals and co-editor, with Beatriz Caiuby Labate, of ten books.

CONTRIBUTOR BIOS

Alexander B. Belser, PhD

Alex Belser is a psychologist and psychedelic researcher in clinical trials at Yale University and a founder of the psychedelic research group at New York University in 2006. He is the Chief Clinical Officer of Cybin. He is a member of Chacruna Institute's Women, Gender Diversity, and Sexual Minorities Working Group.

Anya Ermakova, PhD

Anya Ermakova is a conservation biologist and psychedelic scientist. She researches peyote ecology and conservation in Texas, USA with Prof. Martin Terry. She is part of Chacruna's Council for the Protection of Sacred Plants and a board member of the Cactus Conservation Institute.

Ashleigh Murphy-Beiner, MSc

Ashleigh Murphy-Beiner is a Trainee Clinical Psychologist and Mindfulness Practitioner. She was part of the team working on the Psilocybin for Depression clinical trial at Imperial College London. She has also published research investigating mindfulness and cognitive flexibility in the afterglow period of ayahuasca use.

Belinda Eriacho, MPH

Belinda Eriacho is of Dine' (Navajo) and A:shiwi (Pueblo of Zuni) descent. She is the founder of Kaalogii, which focuses on cultural and traditional teaching and inner healing. She holds degrees in Health Sciences, Technology, and Occupational and Environmental Health.

Bett Williams

Bett Williams is the author of *Girl Walking Backwards.* Her new book, *The Wild Kindness* chronicles cover seven years of psilocybin growing and use, both solitary and in community with queer artists and writers.

Bill Brennan, PhD (C)

Bill Brennan is a psychologist-in-training in New York City. He has trained and taught within the Consciousness Medicine lineage. He is also the co-developer of the EMBARK approach to psychedelic-assisted psychotherapy with Dr. Alex Belser at Cybin.

Celina De Leon, MDiv
Celina De Leon is the director of Circle of Sacred Nature church, a founding member of the UC Berkeley Center for the Science of Psychedelics, adjunct faculty at the Graduate Theological Union, and chair of the board of Chacruna Institute for Psychedelic Plant Medicines.

Charlotte Walsh, MPhil
Charlotte Walsh is a legal academic at Leicester Law School. Her writings and talks challenge the drug laws as they pertain to psychedelics from a liberal, human rights-informed perspective.

Diana Negrín, PhD
Diana Negrín is a geographer, educator and curator. She currently serves on the board of the Wixárika Research Center and teaches at the University of San Francisco and the University of California, Berkeley. She is Associate Director of Chacruna Latinoamérica in Mexico.

Emily Sinclair, PhD (C)
Emily Sinclair is an anthropology PhD (C) candidate studying ayahuasca shamanism in Loreto, Peru. She is a member of Chacruna's Ayahuasca Community Committee and led the Sexual Abuse Awareness initiative.

Erik Davis, PhD
Erik Davis is an independent scholar and lecturer. He is the author, most recently, of Nomad Codes: Adventures in Modern Esoterica. He hosts the weekly podcast Expanding Mind, and earned his PhD from Rice University.

Erika Dyck, PhD
Erika Dyck has a PhD in History, and is currently a Professor at the University of Saskatchewan, and a Canada Research Chair in the History of Health & Social Justice. Erika is also part of Chacruna's Women, Gender Diversity, and Sexual Minorities Working Group, where she hosts the series "Women in the History of Psychedelic Plant Medicines."

Gabriel Amezcua, MA (C)
Gabriel Amezcua is a Mexican anthropologist, conflict mediator and

community builder, specializing in harm reduction, gender studies, and psychedelic therapy. He currently resides in Berlin, where he co-founded the Psychedelic Society Berlin and where he has worked as a consultant for the Open Society Foundations (OSF) and the Multidisciplinary Association for Psychedelic Studies (MAPS).

Geoff Bathje, PhD

Geoff Bathje is a licensed psychologist, full professor, activist, and co-founder of Chicago-based Sana Healing Collective, with additional training in Psychedelic-Assisted Therapies and Research, Sensorimotor Therapy, harm reduction, and community organizing.

Glauber Loures de Assis, PhD

Glauber Loures de Assis has a PhD in Sociology. He is Associate Director of Chacruna Latinoamérica in Brazil and president of Céu da Divina Estrela, a Brazilian Santo Daime church.

Gretel Echazú, PhD

Ana Gretel Echazú Böschemeier has a doctorate degree in Anthropology and a post-doctorate in collective health. She works as an assistant professor at the Federal University from Rio Grande do Norte, Brazil.

Inti García Flores

Inti García Flores is a Mazatec historian from La Salle University, Puebla, and a secondary school teacher in San Mateo Yoloxochitlan. He is also Director of the "Renato García Dorantes" Mazatec Historical Archive.

Jeanna Eichenbaum, LCSW

Jeanna Eichenbaum is a psychotherapist in private practice in San Francisco, and a practitioner of Ketamine Assisted Psychotherapy (KAP) at Healing Realms Center. She specializes in the treatment of trauma and issues impacting the queer community.

Joseph Rhea, JD, PhD

Joseph Rhea is an attorney in Palm Springs, California. Joseph received his doctorate in sociology from Harvard University and works primarily in the area of criminal defense and plant legalization. He is a member of Chacruna's Council for the Protection of Sacred Plants.

Juliana Mulligan, BA

Juliana Mulligan has been an active member of the Ibogaine community for nine years and is currently working on her Masters in Social Work. She runs Inner Vision Ibogaine which supports people in preparation and integration around ibogaine treatment. She is also the Psychedelic Program Coordinator at the Center for Optimal Living. She came to work with ibogaine through her own journey with opioid dependency and incarceration.

Katherine A. Costello, PhD

Katherine A. Costello is a transformational life coach serving the LGBTQIA2S+, poly, and psychedelic communities. She is also an independent scholar working at the intersection of feminist, queer, and transgender theories.

Leopardo Yawa Bane

Leopardo Yawa Bane was born in the Indigenous land Kaxinawá of Rio Jordão, in the state of the Acre, Brazil. He left the Amazon to better learn how to lead his people through our modern times and has been traveling the world leading healing ceremonies for the last 10 years.

Luan Gomes dos Santos de Oliveira

Luan Gomes dos Santos de Oliveira, Anthropologist. Doctorate in Education and professor at the Federal University of Campina Grande - UFCG, Brazil. Topics of interest: Political Ecology, Health and Education.

Marca Cassity, BSN, LMFT

Marca Cassity is a two-spirit trauma therapist specializing in Native American and queer-related trauma. They are trained by MAPS in MDMA-assisted psychotherapy and are part of two MDMA-assisted psychotherapy research studies.

Martin Terry, DVM, PhD

Martin Terry has an inordinate fondness for tackling multifaceted ethnobotanical, ecological, economic, regulatory and conservation problems involving plants, particularly cacti, and most especially those bearing the first name

Lophophora. He and Deirdre Terry and friends are currently attempting to rescue a chunk of the Chihuahuan Desert in the form of a new 501(c)3.

Matthew Baggot, PhD

Matthew Baggott is a data science leader and neuroscientist who has pioneered research on how psychedelics affect people for over two decades. He is founder and CEO of Tactogen, a public benefit corporation, and was previously a director at Genentech.

Mellody Hayes, MD

Mellody Hayes is the CEO of Upgrade My Heart, an innovative wellness company, and a cofounder of Decriminalize Nature and How We Heal. A graduate of Harvard College and UCSF medical school, Dr. Hayes is a leader, writer, and spiritual teacher who is a physician on the MAPS-sponsored FDA Phase 3 clinical trial of MDMA-assisted psychotherapy for Post-Traumatic Stress Disorder.

Monnica T. Williams, PhD, ABPP

Monnica T. Williams is a board-certified clinical psychologist and Associate Professor at the University of Ottawa, where she is the Canada Research Chair in Mental Health Disparities. She has published over 150 articles on ethnic minority mental health, psychopathology research, and psychedelic therapy. She has served on Chacruna's Board of Directors and is currently part of their Racial Equity and Access Committee.

NiCole T. Buchanan, PhD

NiCole T. Buchanan is a Professor of Psychology at Michigan State University, the Clinical Director and Founder of Alliance Psychological Associates, PLLC. in East Lansing, MI, and a member of Chacruna's Racial Equity and Access Committee.

Pietro Benedito, PhD

Pietro Benedito is a sociologist. He specialized in Sociology of Religion through a thesis on Gender and Religion within the Brazilian Ayahuasca Religion Santo Daime.

Rosalía Acosta López

Rosalía Acosta López is a psychologist from La Salle University, Puebla, and has been based in Huautla de Jiménez for 20 years. She currently organizes community reading circles for children in Huautla de Jiménez.

Sarai Piña Alcántara

Sarai Piña Alcántara graduated with a degree in ethnology from the National Faculty of Anthropology and History. She studies social anthropology at CIESAS-CDMX. She has conducted research in the area for 15 years.

Sean Lawler, MFA

Sean Lawlor is a writer and graduate student of Transpersonal Counseling at Naropa University. His writing on psychedelic and healing has appeared in Psychedelics Today, Lucid News, and Chacruna, and his first book on the subject is forthcoming from Sounds True.

UMIYAC

The Union of Indigenous Yagé Doctors of the Colombian Amazon (UMIYAC) is a grassroots organization representing Siona, Cofán, Inga, Kamentsá, and Coreguaje spiritual authorities. Authors include Rubiela Mojomboy, indigenous leader; Ernesto Evanjuanoy, UMIYAC's president; Miguel Evanjuanoy, environmental engineer; Riccardo Vitale, anthropologist, PhD.

Vilmarie Fraguada Narloch, PsyD

Vilmarie Fraguada Narloch is a psychologist and harm reductionist with a certificate in psychedelic-assisted therapy and research, is the director of drug education at SSDP, and co-founder of Sana Healing Collective.

References

WHY BLACK PEOPLE SHOULD EMBRACE PSYCHEDELIC HEALING: RECLAIMING A CULTURAL BIRTHRIGHT

1. Samantha C Holmes, Vanessa C Facemire, and Alexis M DaFonseca, "Expanding criterion a for posttraumatic stress disorder: Considering the deleterious impact of oppression," *Traumatology* 22, no. 4 (2016).
2. Jason B Luoma et al., "A Meta-analysis of Placebo-controlled Trials of Psychedelic-assisted Therapy," *Journal of Psychoactive Drugs* 52, no. 4 (2020).
3. Monnica T Williams et al., "People of color in North America report improvements in racial trauma and mental health symptoms following psychedelic experiences," *Drugs: Education, Prevention and Policy* (2020).
4. Timothy I Michaels et al., "Inclusion of people of color in psychedelic-assisted psychotherapy: A review of the literature," *BMC psychiatry* 18, no. 1 (2018).
5. Michelle Alexander, The new Jim Crow: Mass incarceration in the age of colorblindness (The New Press, 2020); Katherine Beckett, Kris Nyrop, and Lori Pfingst, "Race, drugs, and policing: Understanding disparities in drug delivery arrests," *Criminology* 44, no. 1 (2006).
6. Harold A Abramson, "Lysergic Acid Diethylamide (LSD-25): XXXI. Comparison by questionnaire of psychotomimetic activity of congeners on normal subjects and drug addicts," *Journal of Mental Science* 106, no. 444 (1960).
7. B SMITH et al., "ANNOTATED BIBLIOGRAPHY OF PAPERS FROM THE ADDICTION RESEARCH CENTER, 1935-75," (1978).
8. E.g.: DE Rosenberg et al., "Observations on direct and cross tolerance with LSD and d-amphetamine in man," *Psychopharmacologia* 5, no. 1 (1963).
9. Danny Nemu, "Getting high with the most high: Entheogens in the Old Testament," *Journal of Psychedelic Studies* 3, no. 2 (2019).
10. Zerihun Doda Doffana, "Sacred natural sites, herbal medicine, medicinal plants and their conservation in Sidama, Ethiopia," *Cogent Food & Agriculture* 3, no. 1 (2017), https://doi.org/10.1080/23311932.2017.1365399.
11. Peter Brackenridge, "Ibogaine therapy in the treatment of opiate dependency," *Drugs and Alcohol Today* (2010).
12. Jean-Francois Sobiecki, "Psychoactive ubulawu spiritual medicines and healing dynamics in the initiation process of Southern Bantu diviners," *Journal of Psychoactive drugs* 44, no. 3 (2012).

13. Richard Evans Schultes, A Hofmann, and C Ratsch, *Plants of the Gods: Their Sacred, Healing, and Hallucinogenic Powers* (Rochester: Healing Arts Press, 2010).
14. Jean-Francois Sobiecki, "A preliminary inventory of plants used for psychoactive purposes in southern African healing traditions," *Transactions of the Royal Society of South Africa* 57, no. 1-2 (2002).
15. Mailae Halstead et al., "Ketamine-Assisted Psychotherapy for PTSD Related to Racial Discrimination," *Clinical Case Studies* (2021).
16. Jennifer Dore et al., "Ketamine assisted psychotherapy (KAP): patient demographics, clinical data and outcomes in three large practices administering ketamine with psychotherapy," *Journal of psychoactive drugs* 51, no. 2 (2019).
17. Monnica T. Williams, Sara Reed, and Jamilah George, "Culture and psychedelic psychotherapy: Ethnic and racial themes from three Black women therapists," *Journal of Psychedelic Studies* 4, no. 3 (2021).

MESTRE IRINEU: A BLACK MAN WHO CHANGED THE HISTORY OF AYAHUASCA

1. Paulo Moreira and Edward MacRae, *Eu venho de longe: Mestre Irineu e seus companheiros* (SciELO-EDUFBA, 2011).
2. Moreira and MacRae, *Eu venho de longe: Mestre Irineu e seus companheiros.*
3. Moreira and MacRae, *Eu venho de longe: Mestre Irineu e seus companheiros.*
4. Moreira and MacRae, *Eu venho de longe: Mestre Irineu e seus companheiros.*
5. Beatriz Caiuby Labate, G Assis, and Clancy Cavnar, "A religious battle: Musical dimensions of the Santo Daime diaspora," in *The world ayahuasca diaspora: Reinventions and controversies* (2016).
6. Moreira and MacRae, *Eu venho de longe: Mestre Irineu e seus companheiros.*
7. Beatriz Caiuby Labate and G Assis, "The religion of the forest: Reflections on the international expansion of a Brazilian ayahuasca religion," in *The world ayahuasca diaspora: Reinventions and controversies* (2016).
8. Odemir Raulino da Silva, "Ser Divino," https://nossairmandade.com/hymn.php?hid=1540.
9. Irineu Serra, "Eu Tomo Esta Bebida," *O Cruzeiro*, https://www.nossairmandade.com/hymn/92/EuTomoEstaBebida.
10. Irineu Serra, "Estou Aqui " *O Cruzeiro*, https://www.nossairmandade.com/hymn/301/EstouAqui.

HOW WOMEN HAVE BEEN EXCLUDED FROM THE FIELD OF PSYCHEDELIC SCIENCE

1. Held on November 19, 2018, in San Francisco, California. For more details, see: "Women and Psychedelics Forum," *Chacruna Institute*, 2018, https://chacruna.net/women-and-psychedelics-forum/

WHEN FEMINISM FUNCTIONS AS WHITE SUPREMACY: HOW WHITE FEMINISTS OPPRESS BLACK WOMEN

1. Mariana Ortega, "Being lovingly, knowingly ignorant: White feminism and women of color," *Hypatia* 21, no. 3 (2006).
2. Sarah Jane Brubaker, "Denied, embracing, and resisting medicalization: African American teen mothers' perceptions of formal pregnancy and childbirth care," *Gender & Society* 21, no. 4 (2007).
3. Ann M Dozier et al., "Patterns of postpartum depot medroxyprogesterone administration among low-income mothers," *Journal of Women's Health* 23, no. 3 (2014).
4. James W Collins Jr et al., "Very low birthweight in African American infants: the role of maternal exposure to interpersonal racial discrimination," *American journal of public health* 94, no. 12 (2004).
5. Walter S Gilliam et al., "Do early educators' implicit biases regarding sex and race relate to behavior expectations and recommendations of preschool expulsions and suspensions," *Yale University Child Study Center* 9, no. 28 (2016).
6. Chloë FitzGerald and Samia Hurst, "Implicit bias in healthcare professionals: a systematic review," *BMC Medical Ethics* 18, no. 1 (2017), https://doi.org/10.1186/s12910-017-0179-8
7. Alexandra Minna Stern, "Sterilized in the name of public health: race, immigration, and reproductive control in modern California," *American journal of public health* 95, no. 7 (2005); California. State Auditor, *Sterilization of female inmates: Some inmates were sterilized unlawfully, and safeguards designed to limit occurrences of the procedure failed* (California State Auditor, 2014).
8. Mamta Motwani Accapadi, "When White Women Cry: How White Women's Tears Oppress Women of Color," *College Student Affairs Journal* 26, no. 2 (2007).
9. Robin DiAngelo, "White Fragility," *The International Journal of Critical Pedagogy* 3, no. 3 (2011).

HATE & SOCIAL MEDIA IN PSYCHEDELIC SPACES

1. Shawn Sunil Verma, "Can Psilocybin Make People Feel More Empathy to Nature?," *Chacruna Institute*, 2018, https://chacruna.net/can-psilocybin-people-feel-empathy-nature/.
2. Eddie Jacobs, "What if a Pill Can Change Your Politics or Religious Beliefs?," *Scientific American*, 2020, https://www.scientificamerican.com/article/what-if-a-pill-can-change-your-politics-or-religious-beliefs/.
3. Emily Blatchford, Stephen Bright, and Liam Engel, "Tripping over the other: Could psychedelics increase empathy?," *Journal of Psychedelic Studies JPS* 4, no. 3 (2021), https://doi.org/10.1556/2054.2020.00136.

4. Leor Roseman, "Palestinians, Israelis, and Ayahuasca: Can Psychedelics Promote Reconciliation?" (paper presented at the Breaking Convention, 2019), https://www.youtube.com/watch?app=desktop&v=pH6oL-iyeqs.
5. Monnica T. Williams and Beatriz C. Labate, "Diversity, equity, and access in psychedelic medicine," *Journal of Psychedelic Studies* 4, no. 1 (2020), https://doi.org/10.1556/2054.2019.032.

NEW NARRATIVES WITH PSYCHEDELIC MEDICINE: AN INTERVIEW WITH MELLODY HAYES

1. NiCole T. Buchanan, "13 Steps for Promoting Access and Inclusion in Psychedelic Science," *Chacruna Institute*, 2019, https://chacruna.net/thirteen-steps-for-promoting-access-and-inclusion-in-psychedelic-science-you-say-you-value-diversity-heres-how-to-prove-it/; "People of Color Making a Difference in Psychedelic Healing," *Chacruna Institute*, 2019, https://chacruna.net/people-of-color-making-a-difference-in-psychedelic-healing/.
2. Roland R. Griffiths et al., "Psilocybin produces substantial and sustained decreases in depression and anxiety in patients with life-threatening cancer: A randomized double-blind trial," *Journal of psychopharmacology* 30, no. 12 (2016); Kathryn L. Tucker, "Can the Psychedelic Movement Learn from the Movement for End of Life Liberty?," *Chacruna Institute*, 2019, https://chacruna.net/can-the-psychedelic-movement-learn-from-the-movement-for-end-of-life-liberty/.
3. Jessica Nielson, "Building a Psychedelic Community During the War on Drugs," *Chacruna Institute*, 2019, https://chacruna.net/building-a-psychedelic-community-during-the-war-on-drugs/.
4. Dee Dee Goldpaugh, "Embodied Healing: A Personal Perspective on Resolving Trauma with Psychedelics," *Chacruna Institute*, 2019, https://chacruna.net/embodied-healing-a-personal-perspective-on-resolving-trauma-with-psychedelics/.
5. "Animal Rights Groups Say the Porsolt Swim Test is Unnecessary. But is it?," 2018, https://fbresearch.org/forced-swim-test/.
6. Bruce K. Alexander, "Addiction: The View from Rat Park," 2010, https://www.brucekalexander.com/articles-speeches/rat-park/148-addiction-the-view-from-rat-park.
7. Ismail L. Ali, "A New Era of Psychedelic Policy?," *Chacruna Institute*, 2018, https://chacruna.net/new-era-psychedelic-policy/.

WHY PSYCHEDELIC SCIENCE SHOULD PAY SPEAKERS AND TRAINERS OF COLOR

1. Alnoor Ladha and Martin Kirk, "Psychedelic communities, social justice, and kinship in the Capitalocene," *Kahpi*, 2019, https://kahpi.net/psychedelic-communities-social-justice/.

2. This article frames the comparison as people of color and Whites, but the arguments are easily extended to other marginalized people (e.g., LGBTQIA+ speakers).
3. *The gender wage pay gap and public policy: Briefing paper.*, Institute for Women's Policy Research (2016), http://www.womensfundingnetwork.org/wp-content/uploads/2014/03/The-Gender-Wage-Gap-and-Public-Policy.pdf.
4. Tami Luhby, "5 Disturbing Stats On Black-White Financial Inequality", *Cnnmoney*, 2014, https://money.cnn.com/2014/08/21/news/economy/black-white-inequality/.
5. Kimberly Amadeo, "How To Close The Racial Wealth Gap In The United States", *The Balance*, 2019, https://www.thebalance.com/racial-wealth-gap-in-united-states-4169678.
6. Isis H Settles, NiCole T. Buchanan, and Kristie Dotson, "Scrutinized but not recognized:(In) visibility and hypervisibility experiences of faculty of color," *Journal of Vocational Behavior* 113 (2019).
7. Jeannette Diaz, "A psychological framework for social justice praxis," *The praeger handbook of social justice and psychology: Fundamental issues and special populations* (2014); Linda Tuhiwai Smith, *Decolonizing methodologies: Research and indigenous peoples* (Zed Books Ltd., 2013).
8. MAPS (Multidisciplinary Association for Psychedelic Science), 2019, "LSD-assisted psychotherapy", https://maps.org/research/psilo-lsd.

GUIDELINES AND CONSIDERATIONS FOR WORKING WITH INDIGENOUS PEOPLE IN PSYCHEDELIC SPACES

1. "Native American Rights Fund," n.d., https://www.narf.org/frequently-asked-questions/.
2. Peggy V Beck, Anna Lee Walters, and Nia Francisco, "The Sacred: Ways of Knowledge," *Sources of Life (Tsaile, Ariz.: Navajo Community College Press, 1977)* (1996).
3. Beck, Walters, and Francisco, "The Sacred: Ways of Knowledge."
4. Beck, Walters, and Francisco, "The Sacred: Ways of Knowledge."
5. E.g., C. A. Pouchly, "A narrative review: Arguments for a collaborative approach in mental health between traditional healers and clinicians regarding spiritual beliefs," *Mental Health, Religion & Culture* 15, no. 1 (2012), https://doi.org/10.1080/13674676.2011.553716.
6. E. Duran et al., "International handbook of multigenerational legacies of trauma," in *Healing the American Indian soul wound* (Springer, 1998).
7. D. Bassett, "Our culture is medicine: Perspective of native healers on posttrauma recovery among American Indian and Alaska native patients," *The Permanente Journal* 16, no. 1 (2012), https://doi.org/10.7812/tpp/11-123.
8. Bureau of Indian Affairs, "Mission statement," 2019, https://www.bia.gov/bia.

COLONIAL SHADOWS IN THE PSYCHEDELIC RENAISSANCE

1. Diana Negrín da Silva, ""It is loved and it is defended": Critical Solidarity Across Race and Place," *Antipode* 50, no. 4 (2018), https://doi.org/https://doi.org/10.1111/anti.12396.
2. "Peyote: Territory, Roots and Conflict | Plantas Sagradas en las Américas," (2018). https://www.youtube.com/watch?v=cTbw1DRLQYE.
3. "Closing of Sacred Plants Conference | Plantas Sagradas en las Américas," (2018). https://youtu.be/wFdwoeBvu94.

CULTURAL APPROPRIATION AND MISUSE OF ANCESTRAL YAGÉ MEDICINE

1. The *cabildos* are indigenous political, territorial units recognized by the 1991Colombian Constitution, through Law 21/1991, which approves the adoption of International Labor Organization (ILO) Covenant 169/1989 on the rights of indigenous people.
2. *Taita* is the Kichwa word for father. The term, adopted by several Amazonian Indigenous groups, refers to elders who, after decades of practicing ancestral medicine, have earned the status of spiritual authority. Being called taita is a sign of great prestige.
3. Whites: persons not recognized as belonging to an indigenous ethnic group and who are not members of cabildos or communities.
4. E.g., the case of Mister Edgar Orlando Gaitán, the "neo-shaman" accused in Colombia by the General Prosecutor's Office of having raped and sexually assaulted up to 50 women, some of them underage. See also the case of Mr. A. Varela, of Ayahuasca International. This case gained international notoriety, generating an interesting debate on neo-shamanism and cultural appropriation. One hundred international academics prepared a letter supporting the Cofán people and denouncing the bad practices of Varela's organization.
5. In Colombia, a 2008 Resolution of the Ministry of the Environment and Development recognizes as a natural reserve the Flora Sanctuary of Medicinal Plants, Orito, Ingi Ande, in the Department of Putumayo. The same Resolution indicates that the practice of yagé medicine is a key element of the ancestral cultures of the Inga, Siona, Cofán, Kamentsá, and Coreguaje, and makes reference to the work of Unión de Médicos Indígenas Yageceros de la Amazonia Colombiana (UMIYAC) as the authority of tradition and self-government that works to preserve and revitalize the ancestral cultural practices of the region.
6. Reference to Covenant 169 of the International Labour Organization (ILO) of 1989, adopted in its entirety by the Constitution of Colombia of 1991; United Nations Declaration on the Rights of Indigenous Peoples of 2007; Sentence T-025 of the Constitutional Court of Colombia of 2004 and Resolution 004 of 2009.

7. The general feeling in the organization is that the relationship between academia and the indigenous communities has been historically one of imbalance and disadvantage for the latter.
8. Resolution 004 of 2009, follow-up to Sentence T-025 of 2004 includes the Inga, Kamentsá, Cofán, Coreguaje, and Siona peoples in the list of peoples at risk of physical and cultural extermination. Resolution 004 recognizes the importance of the relationship between territory, traditional institutions and self-government, autonomy and the resilience of indigenous peoples.
9. The People's Ombudsman reports that, between January 1, 2016 and March 31, 2017, 156 social leaders and defenders of human rights were assassinated in Colombia.

THE REVOLUTION WILL NOT BE PSYCHOLOGIZED: PSYCHEDELICS' POTENTIAL FOR SYSTEMIC CHANGE

1. Shantideva, *The Way Of The Bodhisattva* (Boston, MA: Shambhala, 2006).
2. Taylor Lyons and Robin L Carhart-Harris, "Increased nature relatedness and decreased authoritarian political views after psilocybin for treatment-resistant depression," *Journal of Psychopharmacology* 32, no. 7 (2018); Marc Gunther, "Could Psychedelics Heal the World?," *Medium*, January 25, 2020, https://elemental.medium.com/can-psychedelics-heal-the-world-4ea4d5339a89; Jules Peck, "Psychedelics for systems change: could drugs help us save the planet?," *openDemocracy*, 15 February 2020, https://www.opendemocracy.net/en/oureconomy/psychedelics-systems-change-could-drugs-help-us-save-planet/; Paul Ratner, "Scientists find magic mushrooms could help fight fascism," *Big Think*, 28 January, 2018, https://bigthink.com/paul-ratner/scientists-find-magic-mushrooms-could-help-fight-fascism.
3. Philip Cushman, *Constructing the self, constructing America: A cultural history of psychotherapy* (Addison-Wesley/Addison Wesley Longman, 1996).
4. Isaac Prilleltensky, *The morals and politics of psychology: Psychological discourse and the status quo* (Suny Press, 1994).
5. Ignacio Martín-Baró and Ignacio Martín-Baró, *Writings for a liberation psychology* (Harvard University Press, 1994).
6. Francoise Bourzat, "Sacred Mushrooms of the Mazatec Tradition: Transforming the Inner Landscape of the Human Psyche," *Chacruna Institute*, 2019, https://chacruna.net/sacred-mushrooms-of-the-mazatec-tradition-transforming-the-inner-landscape-of-the-human-psyche/; Ben Feinberg, "Conflict and Transformation - Mazatec & Outsiders' Views, Mushroom Use in Huautla," (2017), YouTube video, https://youtu.be/OOJNAqneL10; Richard Katz, *Indigenous healing psychology: Honoring the wisdom of the First Peoples* (Simon and Schuster, 2017).
7. loadedshaman, "Terence McKenna - Evolving Times," (2011), YouTube video, https://youtu.be/7PucjQXO2ko.

CAPITALISM ON PSYCHEDELICS: THE MAINSTREAMING OF AN UNDERGROUND

1. "Statement on Open Science for Psychedelic Medicines and Practices," *Chacruna Institute*, 2018, https://chacruna.net/cooperation-over-competition-statement-on-open-science-for-psychedelic-medicines-and-practices/.
2. Erik Davis, "The Future of Psychedelic Discourse," 2015, https://techgnosis.com/the-future-of-psychedelic-culture/.
3. Erik Davis, "The Bad Shaman Meets the Wayward Doc," *trip*, 2006, http://www.tripzine.com/listing.php?id=650.

PROFITDELIC: A NEW PSYCHEDELIC CONFERENCE TREND

1. Beatriz C Labate, "The Emergence of a New Market: Psychedelic Science Conferences," *Chacruna Institute*, 2020, https://chacruna.net/the-emergence-of-a-new-market-psychedelic-science-conferences/.
2. "We Will Call It Pala," 2020, https://hereandnowstudios.com/we-will-call-it-pala.
3. Leia Friedman, "It's 2020 and White Men Still Dominate Psychedelic Conferences," *Lucid News*, 2020, https://www.lucid.news/men-still-dominate-psychedelic-conferences/.

PSYCHEDELICS ARE QUEER, JUST SAYING

1. Jae Sevelius, "Psychedelic Justice: Disrupting the Cultural Default Mode Network," *Chacruna Institute*, 2018, https://chacruna.net/psychedelic-justice-disrupting-cultural-default-mode-network/.
2. Liam Stack, "Mike Pence And 'Conversion Therapy': A History," *New York Times* (2016). https://www.nytimes.com/2016/11/30/us/politics/mike-pence-and-conversion-therapy-a-history.html.
3. Maureen Groppe, "After Rippon: Searching for clarity on Mike Pence's stance on gay conversion therapy," *IndyStar* (2018). https://www.indystar.com/story/news/politics/2018/02/16/after-rippon-searching-clarity-mike-pences-stance-gay-conversion-therapy/343105002/.

CAN PSYCHEDELICS "CURE" GAY PEOPLE?

1. Jack Drescher and Kenneth J Zucker, *Ex-gay research: Analyzing the Spitzer study and its relation to science, religion, politics, and culture* (Philadelphia, PA: Haworth Press, 2006).
2. Stack, "Mike Pence And 'Conversion Therapy': A History."
3. Groppe, "After Rippon: Searching for clarity on Mike Pence's stance on gay conversion therapy."

4. S Freud, "A letter from Freud. (1951)," *The American Journal of Psychiatry* 107, no. 10.
5. American Psychological Association, "Answers to your questions for a better understanding of sexual orientation and homosexuality", updated October 29, 2008, http://www.apa.org/topics/lgbtq/orientation.
6. Richard Alpert, "Drugs and sexual behavior," *Journal of Sex Research* 5, no. 1 (1969).
7. Stanislav Grof, *Psychology of the future: Lessons from modern consciousness research* (Suny Press, 2019).
8. Robert Masters and Jean Houston, *The varieties of psychedelic experience: The classic guide to the effects of LSD on the human psyche* (Albany, NY: State University of New York Press, 2000).
9. Masters and Houston, *The varieties of psychedelic experience: The classic guide to the effects of LSD on the human psyche.*
10. A Joyce Martin, "The Treatment of Twelve Male Homosexuals with'LSD'(followed by a Detailed Account of One of them who was a Psychopathic Personality)," *Acta Psychotherapeutica et Psychosomatica* (1962).
11. Ronald Sandison, *A century of psychiatry, psychotherapy and group analysis: A search for integration*, vol. 12 (Jessica Kingsley Publishers, 2001).
12. Peter G Stafford and Bonnie Helen Golightly, *LSD: The problem-solving psychedelic* (Award Books New York, 1967).
13. Masters and Houston, *The varieties of psychedelic experience: The classic guide to the effects of LSD on the human psyche*, p39.
14. Frederick Suppe, "Classifying sexual disorders: the diagnostic and statistical manual of the American Psychiatric Association," *Journal of homosexuality* 9, no. 4 (1984).
15. D Denny, "The last time I dropped acid," *Transgender Tapestry* 110 (2006).
16. Denny, "The last time I dropped acid."
17. J Berkowitz, "Word to the mother: How I gave it up on my 30th birthday (I tried it)," *Curve* 6, no. 1 (2008).
18. D Merkur, "A psychoanalytic approach to psychedelic psychotherapy," *Psychedelic medicine: New evidence for hallucinogenic substances as treatments* 2 (2007).
19. Annie Sprinkle, "How psychedelics informed my sex life and sex work," *Sexuality and Culture* 7, no. 2 (2003).

10 CALLS TO ACTION: TOWARD AN LGBTQ-AFFIRMATIVE PSYCHEDELIC THERAPY

1. Clancy Cavanar, "Can Psychedelics "Cure" Gay People?" *Chacruna Institute*, 2018, https://chacruna.net/can-psychedelics-cure-gay-people/; J. Kingsland, "The shameful history of psychedelic gay conversion therapy," *Plastic Brain*, 2019, https://plasticbrainblog.com/2019/05/29/history-psychedelic-gay-conversion-therapy/.

2. Sarah Carr and Helen Spandler, "Hidden from history? A brief modern history of the psychiatric "treatment" of lesbian and bisexual women in England," *The Lancet Psychiatry* 6, no. 4 (2019).
3. JR Ball and Jean J Armstrong, "The use of LSD 25 (D-lysergic acid diethylamide) in the treatment of the sexual perversions," *Canadian Psychiatric Association Journal* 6, no. 4 (1961).
4. Ball and Armstrong, "The use of LSD 25 (D-lysergic acid diethylamide) in the treatment of the sexual perversions."
5. Andrea Ens, " "Wish I would be normal": LSD and Homosexuality at Hollywood Hospital, 1955-1973" (Master's thesis, University of Saskatchewan, 2019).
6. Cavnar, "Can Psychedelics "Cure" Gay People?"
7. "Playboy interview: Timothy Leary," *Playboy*, 1966, https://archive.org/details/playboylearyinteooplayrich/page/n1/mode/2up.
8. Richard Alpert, "LSD and sexuality," *The Psychedelic Review* 10 (1969).
9. Judith M Glassgold et al., "Report of the American Psychological Association task force on appropriate therapeutic responses to sexual orientation," *American Psychological Association* (2009).
10. Rachel A Proujansky and John E Pachankis, "Toward formulating evidence-based principles of LGB-affirmative psychotherapy," *Pragmatic case studies in psychotherapy: PCSP* 10, no. 2 (2014); John E Pachankis, "Uncovering clinical principles and techniques to address minority stress, mental health, and related health risks among gay and bisexual men," *Clinical Psychology: Science and Practice* 21, no. 4 (2014).
11. Pachankis, "Uncovering clinical principles and techniques to address minority stress, mental health, and related health risks among gay and bisexual men."
12. Alexander B Belser et al., "Patient experiences of psilocybin-assisted psychotherapy: an interpretative phenomenological analysis," *Journal of Humanistic Psychology* 57, no. 4 (2017).

DATING MY AYAHUASCA SHAMAN: SEX, POWER, AND CONSENT

1. Emily Sinclair, "Ayahuasca community outreach: Distribution of the guidelines for the awareness of sexual abuse in Iquitos, Peru," *Chacruna Institute*, 2019, https://chacruna.net/ayahuasca-community-outreach-distribution-of-the-guidelines-for-the-awareness-of-sexual-abuse-in-iquitos-peru/.
2. D Peluso, "Ayahuasca's attractions and distractions: Examining sexual seduction in shaman-participant interactions," in *Ayahuasca shamanism in the Amazon and beyond*, ed. Beatriz C Labate and Clancy Cavnar (Oxford, UK: Oxford University Press, 2014), 243.
3. "Ayahuasca Community Guide for the Awareness of Sexual Abuse," *Chacruna Institute*, 2019, https://chacruna.net/community/ayahuasca-community-guide-for-the-awareness-of-sexual-abuse/.

4. "Legal Resources Companion to the Guidelines for the Awareness of Sexual Abuse," *Chacruna Institute*, 2019, https://chacruna.net/legal-resources-companion-to-the-guidelines-for-the-awareness-of-sexual-abuse/.

WHY ONENESS IS NOT INCOMPATIBLE WITH IDENTITY POLITICS

1. Chacruna Institute. (2019, July 1). Queer critique of the psychedelic "mystical experience" [Video]. *YouTube*. https://youtu.be/oRBS57JiTms
2. Delphy, C. (2000). The invention of French feminism: An essential move. *Yale French Studies*, 97, 166–197. doi:10.2307/2903219
3. Foucault, M. (1990). *The history of sexuality. Volume 1: An introduction*. New York City, NY: Vintage Editions/Random House.
4. Irigaray, L. (1985). *This sex which is not one*. (C. Porter & C. Burke, Trans.) Ithaca, NY: Cornell University Press.
5. Labate, B. C., & Buchanan, N. T. (2020, November 5). Hate & social media in psychedelic spaces. *Chacruna.net*. https://chacruna.net/psychedelic-community-social-media-racism/
6. McIntosh, P. (1998). White privilege: Unpacking the invisible knapsack. *Peace and Freedom*, July/August, 10–12.
7. Playboy Magazine. (1966, September). Playboy interview: Timothy Leary. *Playboy Magazine*.

SEXUAL ASSAULT AND GENDER POLITICS IN AYAHUASCA TRADITIONS: A VIEW FROM BRAZIL

1. Suhur. Mash, "PREM BABA, GURU ESPIRITUAL FAZ POSICIONAMENTO SOBRE A ACUSAÇÃO DE ASSÉDIO SEXUAL," (2018). https://youtu.be/Lb8npgeSQtQ.
2. Beatriz C Labate, "Ex-Mestre Geral da UDV solta áudio apoiando o candidato Jair Bolsonaro a presidente do Brasil," 2018, https://www.bialabate.net/news/ex-mestre-geral-da-udv-solta-audio-apoiando-o-candidato-jair-bolsonaro-a-presidente-do-brasil.
3. Bette L Bottoms et al., "Religion-related child maltreatment: A profile of cases encountered by legal and social service agencies," *Behavioral sciences & the law* 33, no. 4 (2015); Marie Keenan, *Child sexual abuse and the Catholic Church: Gender, power, and organizational culture* (Oxford University Press, 2011).

ABUSES AND LACK OF SAFETY IN THE IBOGAINE COMMUNITY

1. "Blessings Of The Forest," n.d., https://www.blessingsoftheforest.org/.

PSYCHEDELIC MASCULINITIES: REFLECTIONS ON POWER, VIOLENCE AND PRIVILEGE

1. Rebekah, "Great Shamanic Deception: Using Ayahuasca as a Conduit for Sexual Fulfillment," *Chacruna Institute*, 2019, https://chacruna.net/great-shamanic-deception-using-ayahuasca-as-a-conduit-for-sexual-fulfilment/; Alhena Caicedo Fernández, "Sexual abuse in the contexts of ritual use of ayahuasca," *Chacruna Institute*, 2018, https://chacruna.net/sexual-abuse-contexts-ritual-use-ayahuasca/.
2. Monica M. Emerich, "Spiritualizing Commodities: Can Media and Market Guide us to an Improved Sustainability?," *University of Illinois Press Blog*, 2011, https://www.press.uillinois.edu/wordpress/spiritualizing-commodities-can-media-and-market-guide-us-to-an-improved-sustainability-by-monica-m-emerich/.

AYAHUASCA COMMUNITY GUIDE FOR THE AWARENESS OF SEXUAL ABUSE

1. "Ayahuasca Community Guide for the Awareness of Sexual Abuse," *Chacruna Institute*, 2019, https://chacruna.net/community/ayahuasca-community-guide-for-the-awareness-of-sexual-abuse/.
2. "Legal Resources Companion to the Guidelines for the Awareness of Sexual Abuse," *Chacruna Institute*, 2019, https://chacruna.net/legal-resources-companion-to-the-guidelines-for-the-awareness-of-sexual-abuse/.

Additional Resources

Fernandez, A. C. "Sexual abuse in the contexts of ritual use of ayahuasca", *Chacruna Institute*, 2018, https://chacruna.net/sexual-abuse-contexts-ritual-use-ayahuasca/. This text is an adaptation of the original: Fernandez, A. C. "Power and legitimacy in the reconfiguration of the yagecero field in Colombia." In *The expanding world ayahuasca diaspora: Appropriation, integration and legislation*, edited by B. C. Labate and C. Cavnar, 199-216. New York City, NY: Routledge 2018.

Peluso, D., "Ayahuasca's attractions and distractions: Examining sexual seduction in shaman-participant interactions, " *Chacruna Institute*, 2018, https://chacruna.net/sexual-seduction-ayahuasca-shaman-participants-interactions/. This text is an adaptation of the original: Peluso, D. "Ayahuasca's attractions and distractions: Examining sexual seduction in shaman-participant interactions." In *Ayahuasca shamanism in the amazon and beyond*, edited by B. C. Labate and C. Cavnar, 321-256. New York City, NY: Oxford University Press, 2014.

Chacruna's Women and Psychedelics Forum, November 19, 2018, CIIS, California: "Women and Psychedelics Forum," *Chacruna Institute*, 2018, https://chacruna.net/women-and-psychedelics-forum/

Women's Visionary Council (2014). "21 Safety Tips for Participating in Ceremonies That Use Psychoactive Substances," https://www.visionarycongress.org/20-safety-tips-for-participating-in-ceremonies-that-use-psychoactive-substances/.

A WORD IN EDGEWISE ABOUT THE SUSTAINABILITY OF PEYOTE

1. *Convention on Biological Diversity*, (Rio de Janeiro: UN, 5 June 1992), *United Nations Treaty Series*, vol. 1760, chap. XXVII, available from https://treaties.un.org/pages/ViewDetails.aspx?src=TREATY&mtdsg_no=XXVII-8&chapter=27
2. Medicinal Plant Specialist Group, *International Standard for Sustainable Wild Collection of Medicinal and Aromatic Plants (ISSC-MAP)*, (2007), https://www.wwf.de/fileadmin/fm-wwf/Publikationen-PDF/Standard_Version1_0.pdf.
3. Tamara Ticktin and Charlie Shackleton, "Harvesting non-timber forest products sustainably: opportunities and challenges," *Non-timber forest products in the global context* (2011).
4. Elinor Ostrom, "A general framework for analyzing sustainability of social-ecological systems," *Science* 325, no. 5939 (2009).
5. Sergio Cristancho and Joanne Vining, "Culturally defined keystone species," *Human Ecology Review* 2004).
6. Martin Terry, "Lophophora williamsii " in *The IUCN Red List of Threatened Species* (2017). https://dx.doi.org/10.2305/IUCN.UK.2017-3.RLTS.T151962A121515326.en.
7. Martin Terry et al., "Limitations to natural production of Lophophora williamsii (Cactaceae) I. Regrowth and survivorship two years post harvest in a South Texas population," *Journal of the Botanical Research Institute of Texas* (2011); Martin Terry et al., "Limitations to natural production of Lophophora williamsii (Cactaceae) II. Effects of repeated harvesting at two-year intervals in a South Texas population," *Journal of the Botanical Research Institute of Texas* (2012); Martin Terry et al., "Limitations to natural production of Lophophora williamsii (Cactaceae) III. Effects of repeated harvesting at two-year intervals for six years in a South Texas (USA) population," *Journal of the Botanical Research Institute of Texas* 8, no. 2 (2014).
8. Anya Ermakova et al., "DENSITIES, PLANT SIZES, AND SPATIAL DISTRIBUTIONS OF SIX WILD POPULATIONS OF LOPHOPHORA WILLIAMSII (CACTACEAE) IN TEXAS, U.S.A," *bioRxiv* (2020), https://doi.org/10.1101/2020.04.03.023515.
9. Anya Ermakova, "Peyote Harvesting Guidelines," *Chacruna Institute*, 2019, https://chacruna.net/peyote-harvesting-guidelines/.

IT'S TIME FOR THE PSYCHEDELIC RENAISSANCE TO JOIN THE HARM REDUCTION MOVEMENT

1. Sarah Mars, "Heroin Addiction Care and Control: the British System 1916 to 1984," *Journal of the Royal Society of Medicine* 96, no. 2 (2003), https://www.ncbi.nlm.nih.gov/pmc/articles/PMC539406/.
2. Nina Lakhani, "Mexico lost its war on drugs 75 years ago, author claims," *The Independant* 2015, https://www.independent.co.uk/news/world/americas/mexico-lost-its-war-drugs-75-years-ago-author-claims-10473796.html.
3. Patt Denning and Jeannie Little, *Practicing harm reduction psychotherapy: An alternative approach to addictions* (Guilford Press, 2011); Patt Denning and Jeannie Little, *Over the influence: The harm reduction guide to controlling your drug and alcohol use* (Guilford Publications, 2017); G Alan Marlatt, Mary E Larimer, and Katie Witkiewitz, *Harm reduction: Pragmatic strategies for managing high-risk behaviors* (Guilford Press, 2011); Andrew Tatarsky, *Harm reduction psychotherapy: A new treatment for drug and alcohol problems* (Jason Aronson, 2007).
4. Jae Sevelius, "Psychedelic Justice: Disrupting the Cultural Default Mode Network," *Chacruna Institute*, 2018, https://chacruna.net/psychedelic-justice-disrupting-cultural-default-mode-network/; David Nickles, "The Dire Need for Systemic Critique Within Psychedelic Communities," *Chacruna Institute*, 2018, https://chacruna.net/dire-need-systemic-critique-within-psychedelics-communities/.
5. Rose Jade, "Integrating Underground Psychedelic Use: A Cautionary Note for Licensed Health Care Providers," *Available at SSRN 3181334* (2018), https://ssrn.com/abstract=3181334.
6. E.g., Geoff J Bathje, "Psychedelic-Assisted Therapy During Prohibition," *Chacruna Institute*, 2018, https://chacruna.net/psychedelic-assisted-therapy-prohibition/.
7. "SSDP Peer Education Program," n.d, https://ssdp.org/justsayknow/.
8. Conant v. Walters, 309 F. 3d 629, No. No. 00-17222 (Court of Appeals, 9th Circuit 2002).
9. P. Janet and F. Raymond, *Névroses et idées fixes .... Par le Dr. Pierre Janet* (F. Alcan, 1904). https://books.google.com.au/books?id=vCsXAAAAYAAJ.
10. American Psychological Association, "Ethical principles of psychologists and code of conduct" (2002 amended effective June 1, 2010, and January 1, 2017). http://www.apa.org/ethics/code/index.html.

WHAT DO PSYCHEDELIC MEDICINE COMPANIES OWE TO THE COMMUNITY?

1. Jonathan Ott, *Pharmacotheon: Entheogenic drugs, their plant sources, and history* (Kennewick, WA: Natural Products Co, 1993).
2. Graham Dutfield, "TK unlimited: The emerging but incoherent international law of traditional knowledge protection," *The Journal of World Intellectual Property* 20, no. 5-6 (2017).

Additional Resources

Dutfield, G. (2017). TK unlimited: The emerging but incoherent international law of traditional knowledge protection. *The Journal of World Intellectual Property, 20*(5–6), 144159. https://doi.org/10.1111/jwip.12085

Emanuel, E. J., Wendler, D., Killen, J., and Grady, C. "What makes clinical research in developing countries ethical? The benchmarks of ethical research." *The Journal of infectious diseases* 189, no. 5, (2004): 930-937. https://doi.org/10.1086/381709

Eyal, N., Lipsitch, M., Bärnighausen, T., and Wikler, D. "Opinion: Risk to study non-participants: A procedural approach." *Proceedings of the National Academy of Sciences* 115, no. 32, (2018): 8051–8053. https://doi.org/10.1073/pnas.1810920115

Margolin, M. and Hartman, S. "A report from the rocky path to legal psychedelics." *Playboy.com*, January 16, 2020. https://www.playboy.com/read/psychedelics-legalization

Henderson, L. A., and Glass, W. J. *LSD: Still with us after all these years.* San Francisco, CA: Jossey-Bass, 2004

Kabil, A. "This Mexican medicine woman hipped America to magic mushrooms, with the help of a bank executive." *Timeline.com*, January 4, 2017. https://timeline.com/with-the-help-of-a-bank-executive-this-mexican-medicine-woman-hipped-america-to-magic-mushrooms-c41f866bbf37

Margolin, M. & Hartman, S. (2020, January 16). A report from the rocky path to legal psychedelics. *Playboy.com.* https://www.playboy.com/read/psychedelics-legalization

Mazzucato, M. "Re-imagining health innovation to deliver public value." *Medium.com*, November 2, 2018. https://medium.com/iipp-blog/why-we-need-to-re-imagine-health-innovation-51f920e36d7e

Ott, J. (1993). Pharmacotheon: Entheogenic drugs, their plant sources, and history. Kennewick, WA: Natural Products Co.

Sabina, M. *Maria Sabina: Her life and her chants.* Transcribed and edited by A. Estrada & H. Munn. Santa Barbara, CA: Ross-Erikson, 1981.

Society of Ethnobiology. "Society of Ethnobiology code of ethics." https://ethnobiology.org/about-society-ethnobiology/ethics

Zelner, B. A. "The pollination approach to producing individual and community wellness." *Multidisciplinary Association for Psychedelic Studies Bulletin* 30 no. 1 (2020): 3437. https://maps.org/news/bulletin/articles/439-bulletin-spring-2020/8131-the-pollination-approach-to-delivering-psychedelic-assisted-mental-healthcare

SACRED RECIPROCITY: SUPPORTING THE ROOTS OF THE PSYCHEDELIC MOVEMENT

1. Christine Hall, "Compass Pathways Secures $80M Series B to Boost Depression Clinical Trial," *Crunchbase News*, 2020, https://news.crunchbase.com/news/compass-pathways-secures-80m-series-b-to-boost-depression-clinical-trial/.

Index

Q

R

S

Y

Z

PSYCHEDELIC JUSTICE